Praise for *Strange Behavior*

"Klawans relates his tales well, easily conveying a great deal of information about evolution, the brain and dealing with brain disorders." —*Scientific American*

"Readers who want the latest information on the brain, a refresher course on how the mind works or an engaging reminder that, despite great advances in diagnostic technology, the brain remains a mystery would be wise to consult Dr. Klawans." —*New Age*

"[Dr. Klawans] was a rare neurologist who easily scaled the imaginary divide between the mind and the brain. He illuminates as he entertains." —Jonathan H. Pincus, M.D., Chief of Neurology, Veteran's Administration Hospital, Washington, D.C., and professor of neurology, Georgetown University Medical Center

"A compelling blend of science and human interest, this book presents gripping ideas about neuroscience." —*Science News*

STRANGE
BEHAVIOR

. . .

ALSO BY HAROLD KLAWANS, M.D.

STRANGE
BEHAVIOR

Tales of Evolutionary Neurology

HAROLD KLAWANS, M.D.

W. W. NORTON & COMPANY

NEW YORK • LONDON

Go Blue

First published as a Norton paperback 2001

For information about permission to reproduce selections from this book,
write to Permissions, W. W. Norton & Company, Inc.,
500 Fifth Avenue, New York, NY 10110

Originally published under the title
Defending the Cavewoman

The text of this book is composed in Sabon with the display set in Cuento
Manufacturing by The Maple-Vail Book Manufacturing Group
Book design by JAM Design

Library of Congress Cataloging-in-Publication Data

Klawans, Harold L.
Defending the cavewoman: and other tales of evolutionary neurology/
Harold Klawans.
p.   cm.
Includes index.
ISBN 0-393-04831-4
1. Neurology Case studies.
2. Brain–Diseases Case studies.
3. Brain–Evolution.
I. Title
RC359.K578   2000
616.8–dc21   99-36704

ISBN 0-393-32184-3 pbk.

W. W. Norton & Company, Inc., 500 Fifth Avenue, New York, N.Y. 10010
www.wwnorton.com

W.W. Norton & Company Ltd., 10 Coptic Street, London WC1A 1PU

1 2 3 4 5 6 7 8 9 0

CONTENTS

·

PREFACE

·

THIS IS NOT a primer on how the brain works. Nor is it a textbook on neurology. I have written both types of books and should know what they are and are not. Yet it is more than just a set of clinical tales about interesting and at times downright peculiar patients who happened to appear in my office in a not-so-random fashion. From the first clinical visit to the final clinical vignette, there has been little that was truly random about the process of my patients, selecting me for their neurologist, and my selecting them for this book. Many of the patients came to my attention specifically because their problems were odd or peculiar. And the tales that make up this collection have been carefully culled form my clinical experiences ranging over the last thirty-five years. They have been chosen not for their peculiarity or strangeness, but to present some aspect of a unified theme.

What, then, is that theme?

In one sense it is so complex that it takes an entire set of clinical stories to elucidate it. In another sense, it is simplicity itself.

We are our genes. And our genes are what they are because of their evolution. Nothing new or odd or peculiar in that. So this book is a neurologist's view of evolution. And what does that entail? How our nervous system has evolved and how that evolution affects who we are as a species, what abilitites we have, what diseases we get, and how we got to be what we are. Our brain is organized the way it is because of the pathway of its evolutionary ascent (or descent, if you prefer). A rare disease called "painful-foot-and-moving-toe syndrome" can only be understood by realizing that we had the same predecessor as dinosaurs and birds and that that ancestor had two sets of "brains." So much more than rare diseases is at issue here, for they can only be understood in terms that are both neurological and evolutionary at the very same time. These evolutionary factors play a determining role in both the abilities our brains acquire and how they acquire them, and in the diseases that we develop and how those diseases do what they do to our function and behavior.

How did we acquire the ability to develop symbolic language? To compose complex music? To read? The goal of this book is more modest than to provide a deep understanding of how the brain works—yet at the same time, equally profound. It more humbly grapples with the "whys" of our brain, not the "hows."

Why are there individuals with otherwise normal brains who can hardly read? Why is it that no one with a normal brain is unable to speak? Why does dyslexia exist? This last question leads to other issues that are really neurological questions, ranging from the profound to the trivial. Is reading related to the way in which our brain functions? Is it really possible to read a movie and watch at the same time? Will the computer revolution produce a change in the way our children's brains function?

Why is it more difficult to learn a second language after the age of sixteen? As a nation we tend to teach languages in a manner that dooms us to failure in the acquisition of a second language. Why have we continued to do this for the last century despite the mounting evidence, known by every intelligent educator, that our approach was bound to fail?

Why did Neanderthal man (and woman) commingle and even cohabit with modern man and yet disappear without a trace?

Why is the myth of "man the hunter" just that, a myth?

Why was it woman who played the key role in the ascent of man?

These are all basic neurological questions whose answers lie in the remarkable workings of the brain and the ways it has evolved. That's what this book is all about.

MANY PEOPLE HAVE played a role in making this book possible. From a practical viewpoint the most important person has been my friend and agent, David Hendin. He was instrumental in guiding me into the structural organization of this book, and despite numerous setbacks he never lost faith in this project. Angela von der Lippe, my editor, also grasped the nature of what I wanted to do, and both encouraged and guided me.

But the deepest gratitude must go to those intellectual mentors of mine who have influenced the way I have come to think about the brain and its functions. They include some world-renowned figures in neurology, as well as others who knew little if any neurology. The list starts with Clark Hopkins, an archeologist who excavated Dura Europas. He taught me how to use my eyes and brain at the same time, to see and classify simultaneously, to appreciate and analyze. It is a skill that I learned at age eighteen and is harder to teach to twenty-five-year-old medical students. My neurological mentors to whom thanks are due include Henry Higman, who taught me how to think about basic science while examining a patient; A. B. Baker; John Garvin; Sigwald Refsum; and my colleagues George Bruyn, Frank Morrell, and Oliver Sacks.

Writing is not easy. Writing about patients in ways that convey both compassion and neurological understanding is even harder. Oliver showed me how that could be done. The success of *The Man Who Mistook His Wife for a Hat* has been more than a model for what I have attempted to do in my tales. It opened doors for my books. Harvey Plotnick of Contemporary Books read it and asked me to write my first collection of tales—not to mimic Oliver, but to write from my own perspective. The result was *Toscanini's Fumble* and now, after a succession of other books, *Strange Behavior*. Oliver is truly the father of us all.

PART ONE

. . .

The Ascent of
Cognitive Function

I

—

DEFENDING THE CAVEWOMAN

·

The Window of Opportunity for Learning

I DIDN'T KNOW THE child's name or if she even had been given one. She was about six years old when I was asked to see her in consultation. She had been admitted to the hospital after being discovered locked in a closet in a dilapidated apartment building that was about to be demolished. The inspector who found her had tried to talk to her, but she had said nothing to him, nor made any sounds or other attempts to communicate with him.

By the time I saw her, we knew a little more. We knew that she was about six years old because the radiologists were able to estimate her age by assessing the apparent age of her bones. Estimates of bone age are based upon the pattern of age-related development of the various growth centers of the bones as well as the age-related closures of certain lines of fusion between bones. The bones of the skull, for instance, are separate at birth. There are even two large openings, or fontanels, between the bones that cover the brain. This arrangement allows the skull to expand as the brain grows. But these openings quickly begin to close. The smaller fontanel is gone within a month or two, and the larger fontanel is fused shut at around a year and a half, almost always by two years. These are estimates and depend upon many factors including nutritional status and the presence of diseases.

The girl's nutritional status was not very good. Her weight was in the fifth percentile for six-year-old girls meaning that 95 percent of all six year olds weighed more than her. Her height was in the tenth percentile. Taken together these observations suggested that she might have been underfed over most of the six years of her life (though she may also have been the child of two short, thin parents). Although undernourished when admitted to the hospital, she had not been starved. And she was clean. True, she had been naked when found, but she was neither soiled nor dirty. Someone—we never learned who—had been doing more than just locking her into a closet.

I was asked to see her in order to answer a very complicated question: why couldn't she speak? The possibilities boiled down to two alternatives. It was the classic dichotomy, nature versus nurture, the oldest question of them all. For Young Girl Doe, the name that had been put on her hospital wristband, this was the issue that had to be addressed first. Was it nature? Did she have a neurological disability that prevented her from speaking? Or was it nurture? Had she just never been exposed to language? With exposure to language, the normal brain and even most abnormal brains will acquire language. Thus what I was looking for was not some subtle neurological anomaly but a significant degree of neurological abnormality that would be sufficient to account for a total inability to say even a single word. Or was her brain relatively normal—that is, within the range of function, where exposure to language at the right age (Young Girl Doe was well within this range) would result in the automatic flowering of language?

I suspected the answer as soon as I walked into the room and said, "Hi."

For she looked up at me from her bed and answered, "Hi."

Her brain had acquired at least one word in less than two days. Actually it had acquired several, including "milk," "TV," and "Lacey," the name of the nurse's aide who had all but adopted her. She also made other sounds, one of which clearly meant "Big Bird," although it was hard for most of the adults to distinguish this from her name for Bert of Bert and Ernie fame. Thank God for *Sesame Street.* It saved us from a long extended neurological workup. Her basic examination was normal other than her language deficit, and she was acquiring language even as I examined her. She loved the little pocket flashlight I used, and laughed each time I shined it into her eyes. By the time I sat down next to her bed to write my consult note, she could say "light," and did so each time she pressed the flashlight's button.

Clearly she could acquire language. That meant that it was far more likely that her inability to speak was the result of nurture, not nature. I was certain of this. As I was that she would quickly overcome her deficits, even though I had never personally observed a child deprived of language to this degree. My certainty came from understanding something about how the brain worked and more specifically about its window of opportunity for the acquisition of language. And by the time Young Girl Doe (who by now was called Lacey) left the hospital two weeks later, I had given so much thought to these issues that I even understood why the entire concept of "man the hunter" was little more than a myth and that the triumph of Man, of *Homo sapiens,* was due entirely to the females of our species. The males did little more than supply the seeds which the females nurtured into modern man. The males' behavior was hardly more than the answer to a basic biological urge. Left to their own devices the males of the species would still be living in caves and scraping the same crude flint blades that they had been scraping for hundreds of thousands of years. It was the females' behavior that made man unique because it led to the development of language.

Lacey immediately brought to my mind the French film *The*

Wild Child (or *L'enfant sauvage*). It was not a movie I would have chosen to see on my own since I hate films that have to be read. And this one was a black-and-white film in French with English subtitles. I had been forced to watch it because one of my best friends, a neuropsychologist named David Garron, had shown it during a party at his home. I was too polite to leave. And then it was too fascinating not to watch—not as a movie but as a scientific document, a veritable *roman à clef* of the formative years of clinical neurology and neuropsychology. It was directed by the legendary French director François Truffaut, who also starred in the role of Dr. Jean Marc Gaspard Itard, the man who first described what is now known as Tourette's syndrome. The movie was based on a true story, one that Itard had experienced and then reported to the scientific community.

The subject of this clinical tale was an adolescent boy who, like Lacey, was without either name or language. He had been known as the Wild Boy of Aveyron because he was found living alone in the woods near Aveyron, France, toward the end of the eighteenth century. Given the name Victor by Itard, he was thought to be about twelve years old when captured. There were no radiologists to estimate the age of his bones, so the estimate was based on the overall maturity of his body, an estimate that is fraught with even more errors than one based on X-rays. He was certainly at least ten but could have been fourteen or fifteen. At the time he was discovered, Victor could neither speak nor understand language. In fact, he appeared not even to possess the slightest notion of using words for the purpose of communication.

This is where Professor Itard entered into Victor's life. Itard was a physician interested in the study of human behavior. He could be considered either a neurologist long before neurologists were invented, or a neuropsychologist even longer before that field emerged, or a cross between the two. (One of the great advantages of living before the field in which you labor has been defined is that your interests and pursuits are not constrained by any artificial borders.) Itard had already published the first case of what later became known as Tourette's syndrome. He was both well known and well respected in his field, whatever that was. He took complete charge of Victor. For over five years, Itard tried to teach Victor to speak, to get him to incorporate even the rudiments of language. But simple words proved to be beyond him. Yet Lacey absorbed words like a sponge. In the short time it took for me to examine her, she learned "light" and used it correctly. She both learned words and understood that they were to be used in communication. After several years of effort, Victor was able to understand only a small number of words and phrases; a few utterances like *lait* (milk) and *oh Dieu* (oh God) were all he ever said, and these he often said incorrectly. By the time I stopped by to see Lacey a second time, she had clearly differentiated between Big Bird and Bert, and could say their names well enough that even casual passersby could tell which *Sesame Street* character she meant. By the end of the five years that Itard and Victor spent together, the Wild Boy, though tamed, never came close to acquiring the use of language.

Why was Lacey able to learn words as if that was precisely what her brain was designed to do while for Victor it was a process as far beyond the capabilities of his brain as Windows 95? It was the same old question: nature versus nurture. True, we do not have all of the details of Victor's case history. While it is possible that he may have been retarded or had some other type of neurological disorder, that is unlikely. Itard was a trained observer of neurological function and would have noted such an abnormality. Furthermore, the young boy had learned to survive on his own in the forest of Aveyron, not exactly rocket science, but something that even some rocket scientists might find very difficult.

Most likely the cause of Victor's absence of language was nurture. Or the lack of it. He had been living on his own in that forest for years, and during that time most likely had not been exposed to language—neither to words nor the concept of words as a form of communication. In Victor this lack of exposure to language before puberty translated into a life during which he would never be able to acquire language despite the best efforts of one of the best minds of his era. True he was without television. He had no *Sesame Street* to watch. But neither did any of his contemporaries, all of whom learned to speak French fluently and even to pronounce it correctly.

Yet Lacey was learning English with a speed that was astounding everyone. Not quite everyone. From the moment I first heard her say "light" and watched her smile as she flicked on the flashlight, I knew she would do very well. Language would be the least of her problems. For speech is a skill that the developing brain is built to acquire. This statement is, of course, all backward. The brain wasn't designed to acquire language: that's not how evolution works. The process of our evolution resulted in a brain that automatically acquires language if that brain is exposed to it. Lacey was exposed to words immediately after her discovery. So had Victor been. She acquired language; Victor never did. What was the difference between them? The ages at which speech was first introduced: Lacy was six, Victor about twice her age. The initial acquisition of language can occur only while the brain is still developing, reaching its full potential. Once that process is complete, the window of opportunity for learning a language as a form of symbolic communication is lost.

To realize why such windows of opportunity exist within the human brain, it is necessary to understand how our brain got to be the way it is and how it actually learns and acquires skills. The human brain did not just appear fully developed within our skulls. It evolved as part of the process of classical Darwinian evolution.

The most surprising element of this "ascent of man" is that the absolute increase in both the size and the complexity that characterizes the human brain has been achieved with remarkably little genetic change. There is an embarrassingly close similarity between our genetic makeup and that of the gorilla or the chimpanzee. More than that, the total amount of genetic information coded in the double helixes of DNA has remained fairly constant throughout all of mammalian evolution, whether shrews or kangaroos or dolphins or humans. It is thought that there are about 1 million separate genes. They are divided up into a different number of chromosomes in different species, but the total number of genes is relatively stable, pretty much the same in mouse as in human. In all humans the number is, of course, identical, though the number of *active* genes is far less than 1 million. It is closer to half that since 40 percent or more of all chromosomal DNA appears to be redundant and plays no active role in development at all. That is, half of our genes have not evolved through evolution.

The best estimates suggest that about ten thousand genes, which is 1 percent of the total gene pool (or approximately 2 percent of the active gene pool), play an active part in the design and construction of the brain and the rest of the nervous system. This is true for humans and chimpanzees and walruses, even our pet gerbils.

This number seems more than adequate for the gerbil or the ordinary house cat, or maybe even a chimpanzee. But for humans? Our brains are made up of about 10 billion cells, and the size and complexity does not end there. There are 10^{14} synapses, or active connections between nerve cells, where messages can be relayed or interrupted. That is one hundred trillion. How can a mere ten thousand genes manage to control so many synapses? How can these relatively few genes do so much more for us than they do for other species?

The fact is that most of what these genes do for us and our brains is not that different from what they do in other species. Any survey of comparative anatomy of the nervous systems of mammals resoundingly supports that conclusion. The major structures are all the same whether the brain belongs to a sheep or a human. The visual cortex is always in the back of the brain. Sheep have the same thalamus buried within the two hemispheres, the same hypothalamus integrating the brain and the endocrine system. Most of the major pathways are the same. The motor cortex is at the back end of the frontal lobe in all vertebrate species. The same pathway of neuron fibers descends through the spinal cord to control the muscles on the opposite half of the body. That crossing is as universal as the dependence of muscles on the brain. In essence, the hard wiring laid down by the genes is pretty much the same no matter what the species, so the general structure is far more similar than dissimilar. Yet it is only in humans that the brain keeps developing and growing after birth.

If the hard wiring and the basic structure of the human brain are so similar to those of other species, why do our brains function so differently? Because our evolution and how we function have not entirely been the product of biological evolution, our genetic heritage; that is, the "hard wiring" in our brains is not the end of the story. Adding to the complexity of our brain are social, cultural, and environmental changes. Our genetic coding allows the brain to grow and develop while interacting with the environment. What separates us is that window (remember those spaces in a newborn's skull?), which give the brain the opportunity to grow and learn—for example, to acquire language.

Human infants are underdeveloped and helpless at birth and remain far more dependent for far longer than the offspring of any other species. We are born with an immature, almost embryonic brain, which continues to grow and evolve in relation to its environment to a degree and for a duration of time not found in any other species.

The brains of most other species are fully formed by birth: even the brains of the other primates continue to grow only for a brief, early postnatal period. The brains of humans continue to grow at rapid fetal growth rates long after birth. That is why final closure of our skulls occurs far later than in any other animal. This process of brain development, some of which takes place after the skull closes, extends for many years. The duration is different in different systems of the brain and in some even continues into what we consider adult life. This post-birth prolonged development of the human brain is often referred to as the juvenilization of the brain. The chimpanzee has a gestation period of seven and a half months, which is a month and a half shorter than that of the average human. The chimp, along with its brain, reaches adulthood at nine months of age, far more rapidly than even the most precocious of humans. The newborn chimp can hold its head up within two weeks of birth, while the average human baby takes ten times that long (twenty weeks) to accomplish this same feat. The chimp, having leapt ahead in development, increases the gap, walking by the end of the fourth week of life. Our biped offspring do not accomplish this until the chimp is already cavorting.

At birth, the human brain is only about one quarter of its eventual adult size and weight. The neonatal (newborn) chimp brain is approximately 350 cubic centimeters. As an adult, the size will reach 450 centimeters, an increase of 100 cubic centimeters, a bit over three ounces, and a total expansion of some 28 percent. The newborn baby, in contrast, has a brain capacity similar to that of the newborn chimp—about 350 cubic centimeters. That is where the similarity ends. While the chimp is out being a chimp, the human brain just keeps on growing, reaching a volume of approximately 1,400 cubic centimeters, or four times the average size at birth. This represents an increase of 300 percent, well over ten times the increase managed by our close relatives the chimpanzees. In other words, most of the human brain develops after birth. All of that growth occurs while the brain is functioning at some level, and the vast majority of these brain functions directly relate to the environment. This means that environmental influences can help shape all postnatal development. It is during this prolonged period of dependency, of growth and development of the brain, that the brain is most plastic and thereby most susceptible to environmental influence. It is not just the ten thousand genes that figure out how all those synapses are to interact. The environment helps write the software. It is also during this period that most environmentally dependent skills are acquired by the brain. In essence, then, the brain of the chimp and all other newborns of all other species develops almost entirely within the womb, while the human brain develops primarily outside the womb. (The dolphin actually comes closest to man in this regard, its brain doubling in size after birth.) Man's brain is enriched by environmental input, while the chimp's brain, in contrast, could be said to be environmentally deprived. And no intentionally added postnatal enrichment can ever really close this gap.

Man, during his ascent, must have been selected for such postnatal development; in other words, those few genetic differences which make us *Homo sapiens* must relate primarily to prolonged or extended postnatal expansion and development of the brain. The limiting factor for the size of the head at birth is the size of the adult female pelvis. Dolphins solved this by making their pelvis vestigial, so there is no pelvic limitation on head size at birth. Humans solved the problem by transferring most of the development of the brain to the postnatal period. The sequence of human evolution has followed this order: First came upright posture and bipedal gait without any change in the size of the pelvis and thereby no change in the volume of the newborn brain. (That still hasn't changed: the human brain is about the same size at birth as that of the chimp.) Then came the postnatal development of the brain. But what allowed this to be selected for over the last several million years?

The period of dependency in humans is not brief. It is one of the drawbacks of juvenilization. Neither the infant nor the young child can survive on its own. This lasts many years. Until early adolescence, the human left to his or her own resources would find life all but impossible. This dependency is both a result of the lack of learned functions by the brain, and a sign that the process of acquiring such adaptive skills is still proceeding. The human brain is distinguished from the brains of other species by its postnatal capacity for learning, its apparent plasticity, but there are limits. There are critical periods or windows of opportunity for learning. Certain skills can only be acquired at specific times. If a skill is not acquired during that critical period, then the acquisition of that skill in later life will be harder, if not impossible. Nothing typifies this more than the acquisition of language.

Man is not unique in having such windows of opportunity that are open only for specific parts of the life cycle. Birds learn their specific songs by imitating the songs of others of their kind. To do this, almost every species of birds must hear these songs in the first couple months of their lives. If the songs are not heard during this critical period, they are never learned. Birds deprived of this input remain songless. The one exception to this rule is canaries. It appears that each season canaries can learn new songs. It is almost as if they can recapture their youth. This annual rebirth of a critical period for learning is accompanied by an annual crop of new neurons which makes the acquisition possible.

Human infants acquire a bewildering number of different skills as their brains mature. They learn to sit up, to stand, to crawl, to walk. None of these skills require any teaching. They do not even require any external input. These abilities are not grounded in teaching or even based on mimicry. The parents walk in one morning and the baby is standing up in the crib. A blind infant masters these skills, as does a deaf or neglected infant. It is as if the acquisition of these skills is built into the nervous system. The brain at birth cannot direct these activities, but as it matures it always acquires these functions, and when such abilities are acquired, they automatically are performed. It's called maturation.

Other potential skills are there but depend on environmental input. The ability to acquire songs is ingrained genetically within the brains of birds. This ability is thus waiting to be activated by the environment. The specific song a specific brain will acquire depends on the environmental input. The bird's brain can only acquire a song that it hears. This is analogous to the human acquisition of language. The ability to learn language is genetically encoded in our brains. Given an environmental exposure to language, a normal brain will always acquire language. If a brain gets exposure and does not develop language, then that brain is abnormal.

What specific language a particular brain will learn depends on which language or languages the brain is exposed to during its critical period. So Victor should have learned French, while Lacey should have learned English, had they been merely given the exposure. Children acquire language with very little assistance from anyone else. It is primarily self-taught, almost like learning to walk, the way Lacey learned to say "light" and "Big Bird" and "Bert." Our brains appear to be innately equipped with systems that are able to acquire language. But this innate capability is both governed and limited by the maturation of the brain.

As the human brain matures and acquires specific self-taught "hard-wired" skills, it passes through different stages in its openness to acquiring language. When the child is one year old and capable of standing up alone, that child is also able to duplicate a few syllables and can also understand some words. Six months later he or she is creeping backward and downstairs and can walk forward. The child now has a repertoire of anywhere from a few to fifty or more words. These are not just sounds that are echoed but words that are used and understood. At this stage they are only single words, not phrases. By two years of age, the child is running, falling fairly frequently, but nonetheless running. He or she now uses a greater number of words and uses them as part of short phrases. Babbling, which had begun at about six months when the child began sitting up, disappears. All of these changes, hopefully, are occurring while the child is in a protective environment which supplies language that the growing and developing brain responds to and integrates into its pattern of growth and development. And so it goes until age four, by which time language has become well established.

The process is the same for all children assuming lack of disease; it is the language that differs from culture to culture. Those differences are found not merely in vocabulary. There is great variety among languages in how phrases and sentences, and even the simplest thoughts beyond simple nouns and verbs, are organized. All of these become ingrained. A brain that acquires language and then uses it for internal thought processes may give up the use of spoken language because of deprivation, but the language of internal thought is never lost.

And no matter what culture the human infant is raised in, no matter what language he is exposed to, acquisition of language can only occur during a critical period of development if it is ever to be acquired at all. For language this critical period, or window of opportunity, appears to end around early adolescence.

Victor was the original basis for our understanding that such a critical period exists for the acquisition of language. Since then, other far better documented instances of children deprived of environmental language input have been studied and reported on, all of which seem to demonstrate this same point. A girl dubbed "Genie" is one of the best known. From the age of twenty months she had been kept in an isolated room and deprived of virtually all human contact. By that time she was well into her window of opportunity and should have been able to say a number of words and understand many more. She should have been able to repeat words, know how to use them and may even have begun to string them together into short, but meaningful phrases. Her enforced isolation and linguistic deprivation was continued until she reached the age of thirteen and had already passed puberty. Her brain had reached adult size without the benefit of environmental language. No Big Bird was talking to her. Nor was anyone else. Her obviously schizophrenic father treated her like an animal. He even barked at her instead of talking to her.

When Genie was finally rescued, she was totally without language. Like Victor and Lacey, she could neither speak nor understand. Whatever she may have learned early in her life had been lost. She could not say anything. It was at this point that language exposure and instruction were initiated. Genie did far better than Victor. She did learn to comprehend language reasonably well, but her speech lagged far behind and she never mastered even the rudiments of grammar. According to her mother, she had learned single words prior to her incarceration. So during her "critical period" she had already started to learn language. She had an advantage over Victor in that some key elements of her language learning after puberty actually represented relearning. As a result Genie was able to reattain at least a fair measure of comprehension.

If Victor is the original model of what happens when children are environmentally deprived during their language window, then Genie teaches us that the acquisition process isn't all-or-nothing. Genie was able to learn, though her achievement level was poor. She filled the gap between Victor and Lacey in our understanding of the neurophysiological basis of acquiring language. She shed light on the parameters for how the brain learns language. She showed us that the window of opportunity is not an all-or-nothing proposition. But no matter what the "mechanics" are, the process is bound by a critical period, a window of opportunity.

One can argue that Victor's and Genie's stories are nothing more than peculiar aberrations. Such children were deprived of far more than just speech; both were socially and emotionally deprived as well. Are they the only evidence we have for a critical period during which speech and language must be acquired?

Of course not. Victor was merely the starting place for our understanding. The best support for the concept of a critical period comes from what neurologists have learned by studying patients who have lost the ability to use language, a condition neurologists call aphasia. The sudden onset of aphasia is most commonly caused by stroke. While we commonly think of strokes as a disorder of the older population, the fact is that strokes can occur at any age (though there are different causes at different ages), and they cause the same problems in three year olds as they do in that three year old's grandparent. However, though the nature of the neurological deficit is pretty much the same, what happens after the stroke is not. Our observations on the nature of recovery from aphasia have contributed to our notion of a critical period for speech acquisition, which in turn remains pivotal to our understanding that critical periods are part and parcel of normal brain function and development.

The basic neuroanatomy of loss of acquired speech can be summarized as some general rules which I have derived from generations of neurological observation:

Rule One: The sudden loss of speech implies that something has gone amiss in one hemisphere. We learned about this hemisphere by studying aphasia; therefore we learned "backward" that one hemisphere dominates our ability to speak.

Rule Two: If a patient is aphasic, that patient has a lesion in his or her dominant hemisphere for speech.

Rule Three: In right-handers the dominant hemisphere is invariably the left hemisphere. So if a right-hander is aphasic, he has a problem in his left hemisphere. In left-handers, the situation is not as clear, even though in most left-handers the left hemisphere is dominant for speech.

Rule Four: Not all aphasias are identical; if a patient has more trouble speaking than understanding, the lesion is toward the front of the dominant hemisphere. If the patient has more trouble understanding than speaking, the lesion is toward the back of the hemisphere.

Rule Five: All other theories concerning the localization of speech within the brain are just talk.

What does aphasia have to do with how our brain has evolved to learn language?

When a patient suffers a sudden loss of speech, we know there is an injury to the specific areas of the left hemisphere. Speech is lost. Is that the end of it? That depends on the age of the patient. Hence, aphasia gives us a view into the window of opportunity for learning or relearning speech. Every neurologist knows this from treating brain-injured patients. In childhood, traumas are often the result of inflammation of the cartid artery that supplies blood to one of the cerebral hemispheres. This results in a significant injury of that hemisphere and a hemiplegia, paralysis of the limbs on the opposite side of the body (also called "infantile hemiplegia"). The child very well might not recover from the paralysis, since the hemisphere that contains movement from birth—or even before—is hard-wired. If the dominant hemisphere for speech is affected by this loss of blood supply, the child also develops aphasia and can no longer speak.

If the stroke occurs at age three or four, speech becomes severely impaired, but after a brief period of time, normal speech is rapidly, and almost invariably fully, reacquired. But it is not reacquired by the injured dominant hemisphere. It now becomes housed in what had been the nondominant hemisphere for speech. During normal development, the left hemisphere is selected to become dominant. (At the risk of repeating myself, in left-handers either hemisphere can be selected, but more often the left is chosen.) In a very young child, however, this selection is not yet rigidly set, so that if the left hemisphere is injured, switching dominance is carried out almost without skipping a beat. In fact, once recovery begins, these brain-injured children pass their language milestones at a remarkably accelerated rate until they catch up to exactly where they should be. They then move on as if nothing had ever happened to them. The other side of the brain not only does the job; it does it every bit as well as the first side ever did. So both sides must originally have had equal structural and functional potential as far as the control of speech.

But the ability to switch hemispheres for speech dominance does not go on forever. If it did, adults with strokes would all recover their speech, which we know too well is not what happens. Most children recover speech fully if they are struck before the age of nine. Puberty is the turning point. That is what we learned from Genie. Children who become aphasic between nine and their midteens fall in between. They rarely fully reacquire speech, but they recover more than adults do. By the age of fifteen or sixteen, the prognosis for recovery from an acute aphasia is the same as for adults. Recovery from aphasia is not impossible in adults. But the recovery that occurs is quite different. Most of it occurs rapidly and is due neither to adaptation by the brain in finding a new area on the opposite hemisphere to assume the function of the injured speech region nor to a process of relearning. Rather, it represents healing and "shrinkage" of the initial stroke. Such rapid recovery suggests that the initial "loss" was due to loss of function of areas of the brain that were partially injured but not permanently destroyed. The symptoms remaining after a few weeks tend to be permanent.

So much for the neurologic observations on the windows of learning. But we also have important observations from other fields of study. In fact, some elements of support for the notion of a critical period for language acquisition is well within the experience of almost every American, whether or not that person has ever heard of a child with infantile hemiplegia, or run across a Genie locked in the basement or a Victor wandering the back woods subsisting on his own in the wild. We have all been able to observe the struggles of normal brains to acquire new language skills. We have seen this in parents or grandparents of our own or of our friends, and in wave after wave, generation after generation, of immigrants. We ourselves may have struggled to acquire a new "first" language. The original language of immigrants may change, but the brain still thinks in the original language.

All of us are aware of the problems faced in learning a second language. What adult tourist hasn't returned from France amazed that four-year-old French children have mastered the skill of speaking French complete with correct accents, one that has eluded the adult. Second languages are far easier to acquire during childhood than during adolescence or adulthood, and no matter how much we study the process of learning language, the results will be the same.

Currently, researchers are observing Chinese and Korean immigrants' progress in acquiring English. And guess what? Children learned English quickly and correctly, with no accents at all, right up to the age of puberty. The major element of teaching is immersion into an environment in which English is spoken.

People can and do learn a second language after puberty and even during adult life. But it takes considerable effort, far more than it does prior to puberty. Furthermore, a second language learned after puberty is always that—a second language grafted onto the first rather than a natural language fully and easily acquired.

There is nothing new in this. My grandparents had accents. My father learned a second language quite easily as a child because his parents spoke to each other in that language. It was in his environment, and so his brain acquired it. No formal teaching was required, no second language classes, merely exposure.

What bothers me is that despite neurological studies, clinical observation, and personal experience, which all teach us that acquiring language is directly related to a window of opportunity in a child's development, the teaching of a second language in our schools is not seriously begun until high school, precisely the wrong time to start.

LANGUAGE IS NOT the only system within the brain that is influenced by the environment and for which there is a critical period during which this environmental influence can be expressed. Such is the rule for brain functions, not the exception. That is the way the juvenilized human brain evolves into the adult human brain. It develops in conjunction with environmental influences (nurturing), and without such influences the human brain does not develop normally. This general rule derives its best support from the work of two American neuroscientists, David H. Hubel and Torsten N. Wiesel, who shared the 1981 Nobel Prize for physiology and medicine. Their work did not involve speech but vision. The visual system is truly hard-wired. It is set at birth. If it is injured, no other part of the brain can take over. And if the brain is not injured, it works automatically. So we thought.

The visual cortex of cats and monkeys, and it is assumed also of humans, contain neurons that respond selectively to specific features of the environment such as color or spatial relationships. This selectivity is always there in the normal brain, but it's not as fully automatic as it seems. It depends entirely on the environment. This is what Hubel and Wiesel demonstrated. When a kitten was reared in an environment made up entirely of vertical stripes, then the neurons which responded to visual and spatial orientation "learned" to respond primarily to vertical lines. When horizontal lines were introduced into the cats' environment late in life, these neurons did not respond. In a sense, their brains were not able to see these lines. The "bias" of brain cells to preferentially respond to certain lines and not to others could only be acquired during a critical period of growth in newborn kittens and is an example of a selective rather than an instructive process.

I should step back here and explain that there are two opposing views concerning how the brain learns acquired skills; one is referred to as instructive or constructive, the other selective. The instructive model is the older view: networks of cells are "instructed" by experiences to form certain synapses (pathways) which, once formed and reinforced, become permanent. In this view, the window of learning closes when the nerve cells, either through age or through injury, can no longer reinforce the pathways.

The selective process works just the opposite way. All the kitten's pathways for seeing color and observing spatial relationships exist at birth, waiting to be used. As the juvenilized brain matures, those pathways that do not get used (selected) are eliminated: they atrophy and disappear forever. Most scientists agree that the selective process is most probably the way our brain learns. This process is attractive. It means the more exposure the brain receives, especially at critical times, the more it learns: it can add on languages or colors or sounds, even if there is no formal instruction. But if the brain is not exposed to a nurturing environment, the pathways close down—end of ballgame. In other words, the kitten's system was initially capable of responding to both vertical and horizontal lines, but the artificial limitation of exposure to only vertical lines resulted in vertical bias. This limited the neurophysiological function of the brain. The cats learned the "language" of only vertical lines.

So, like the cats that were never exposed to horizontal lines in early life, Victor had never had language in his environment. The cats never learned to see horizontal lines, and Victor never learned language.

But Lacey did. Her window of opportunity had not passed her by. Her brain could still acquire language, and it did. Not with a vengeance, but with the normal skills of a normal brain interacting with its environment. No teaching was required. Just input. Just an environment with more than vertical lines. And Genie seems to have been rescued just in time to get her part way there.

How did humans get this way? How did language develop? In order to understand we have to shed a few myths ingrained in our culture. We have to turn our attention to the cavewoman. God created man in his image. At least that is the tradition according to Western religious beliefs. That tradition developed and thrived within male-dominated cultures. Our concept of the biological ascent of man arose in that very same cultural milieu. In the century and a half since the publication of Darwin's *Origin of Species,* white male anthropologists and paleontologists designed a blueprint of human evolution in which the success of hunting was viewed as the primary factor in the expansion and dominance of humankind.

According to this conceptualization, "man the hunter" developed better and better tools for hunting prey and as a direct result gained supremacy over the beasts. It was these tools—designed by males, used by males, in the male occupation of hunting—that gave man his great biological advantage in the struggle to survive. Caveman and the wooly mammoth had both struggled to survive the Ice Age. Man had made it. So had the wooly mammoths. But then man devised a secret weapon and bingo, no more wooly mammoths. By making better tools, man changed the playing field. He then pulled woman along behind him, perhaps not by her hair, but certainly not as an equal partner in the ascent. This model was created by men, about men, for men. It proclaims that the success men achieved with their newly created tools resulted in the cultural advances that allowed the Egyptians to build the pyramids and Homer to compose the *Iliad*. Once that was accomplished, the flowering of a literate culture was just around the corner. We are asked to believe that all of this came about because males figured out how to make better flint blades and spears.

The problem is that this "flow diagram" just doesn't stand up to rigorous scrutiny. The fact is that hunting prey for survival is fairly easy. Lions do it quite well. They also teach their cubs how to do it equally successfully. And, of course, lions still can't talk and have not yet developed a literate society. Wolves hunt in coordinated packs and do it quite successfully, generation after generation. Last time I checked they too were without language.

While improved hunting implements could assure a better supply of food, and therefore a decrease in infant mortality (the key to true Darwinian biological superiority), it is difficult to ascribe to such a technological advance any changes other than a mere increase in numbers. Certainly not man's cultural explosion, nor the development of language.

So if it wasn't "man the hunter" who was responsible for the explosive biological advantage of modern humans, what was?

Our advantages over other species are most probably due to the development of a complex language. And women are far more likely to have played the more significant role in this than men. Women were the ones who did the tough job: raising the juvenilized children in caves or any other environment and teaching those children what they needed to know to survive in the world while they were still dependent, weak and slow. Teaching survival to the juvenilized infant depended on language. Language gave the human the distinct advantage for survival. And over a million years or two, the result was the evolution of brains selected for acquisition of language and other skills during the period of prolonged juvenilization.

How long did that take?

A million years here. A million years there. Long enough to be talking about real time. But it was an evolutionary process. Language evolved as mothers talked to children who repeated those words and developed the entire concept of symbolic language. Not merely language as "a means of communication," of screaming out a warning. Many primates can do that. Even an aphasic adult who is all but mute, who cannot say his name or even recognize it, can swear and shout out a warning. Such "speech" is similar to the sudden, explosive warnings of other species, but it is not symbolic language. And it is symbolic language that mothers and children initiated and developed. If the growing child didn't learn enough to survive until he could throw a spear, he would never do very much hunting. Children had to acquire language when young, and they had to learn it from their mothers. The mothers had to have developed language beyond the simple noises needed to hunt. So, though few defend the Cavewoman, we all speak our mother's tongue.

And Lacey speaks her mother's tongue. In fact, she has turned out to be very proficient at languages. She knows English and French, and has developed beyond both Genie and Victor. She also knows Spanish and is learning Polish, both very useful in her job as a nurse in the Critical Care Unit of her home city's largest hospital. What fascinates me about this is that, as a child, Lacey's challenge was to learn speech—the acquired *Homo sapiens* skill for which we can thank our original mothers. As an adult, this experience drove her to one of the nurturing professions. Originally, men hunted while women nurtured, including bringing that juvenilized brain up to speed, so infants could acquire speech. Lacey, who taught me so much about this complex human asset, has gone back to the source—the care and feeding of the infant and her brain. Which brings us back to the Cavewoman.

2

—

A LUCY OF MY VERY OWN

·

Locating Handedness and Speech

Lucy is the name that Donald Johanson, a famous searcher for lost human ancestors, gave to the skeleton he found of one of the earliest species of hominids and which has been given a name of its own, *Australopithecus afarensis*. In just a few years, "Lucy" became the most famous name in all of human paleontology. She was discovered in a remote gorge in Ethiopia, and Johanson had gone there specifically to hunt for her or someone like her. This young woman had lived about 3,500,000 years ago and still had many apelike features, including short legs, a small apelike cranium, and apelike dentition. She was, however, fully capable of bipedal movement. She walked on two legs. This was the important development, for it freed up the hands for far better things than just mobility. Her bipedal gait and the various anatomical changes that made it possible came as no surprise to scientists. In fact, Lucy was the fourth such discovery of an early hominid—three sets of footprints had already been discovered by Mary Leakey at different sites in Tanzania.

The skeleton, which was about 40 percent complete, quickly gained a place of honor in the ascent of humans. She was, in a way, the proverbial "missing link," that long-sought example of an early human ancestor that would bridge the gap between our ancestors and their predecessors. She was not the link between us and the now-living apes, as this term is often interpreted, but a more recent link in the chain of ascent of prehominids that earlier branched off into humans but also gave rise to the various species of chimpanzees, to whom we are so closely related. These very early predecessors of ours had low, apelike cranial vaults, apelike teeth, and short legs. They were still quadrupeds, although they could move bipedally when necessary, just like other chimpanzees. But Lucy and her relatives had crossed the line: they walked upright. Lucy thus answered a question as old as the study of the evolution of man. Which came first, the big brain or bipedalism?

Bipedalism came first. That is what Lucy taught us. Which but not why. It was another, far younger woman, also named Lucy, who first initiated me into my own neurological attempt to consider this issue.

I first met Lucy Arft in 1969. I was not worried about anthropological speculations then. Neither was she nor her very worried mother. The two of them had traveled from Ann Arbor, Michigan, to Chicago to consult with me. They had been told that Lucy most likely had a brain tumor. If someone had told me that my nineteen-year-old daughter had a brain tumor, I too might go looking for a different diagnosis.

Lucy Arft was an honors student at the University of Michigan. She had been healthy, well adjusted with no particular problems, and then for no apparent reason started having sudden unexplained episodes of unconsciousness. These each began with the sensation of a peculiar odor which she could not quite describe. The smell was more like burnt coffee than anything else, but it was not really like any burnt coffee she could actually recall having smelled. Each time the odor was exactly the same.

"Does anyone else smell it?" I asked.

The answer was no.

"Do they smell anything at all?"

The answer was again negative.

That meant this burnt-coffee-like odor was a perception being generated entirely within the brain itself. It was not a misperception of an external stimulus but a result of the spontaneous activity of cells within Lucy's brain. Such abnormal activities always point to cells that are not functioning normally. And it is a short leap doctors make from abnormal function to abnormal structure. If the cells that normally received or interpreted smells were producing smells of their own, perhaps a tumor in that part of the brain was causing these cells to do that.

Each time the smell occurred, Lucy's perception of it lasted for only a few seconds and was invariably followed by smacking of her lips. She was not aware of these movements. They had been seen by other people who were with her—a roommate, her boyfriend, and mother. None of them had smelled the odor, but they would see her smacking her lips and staring at them vacantly. She had no abnormal movements other than lip smacking. But it invariably initiated a period of unconsciousness during which she could not be aroused. Once the episode began, she would stand motionless, looking out through sightless eyes, and neither answer if anyone talked to her nor respond if someone touched her. It was as if she were absent from her brain, almost as if she were no longer there.

"For how long?"

"A couple of minutes. My roommate timed one."

She recalled none of this; it had all been told to her by others. After each one of these "absence attacks," she would wake up feeling absolutely exhausted and would sleep for twenty or thirty minutes and feel just like herself again except that she had no recollection of what had happened to her. Each time there was a new hole in her memory.

The neurologist in Michigan who first evaluated Lucy had correctly diagnosed her spells as psychomotor seizures, a form of epilepsy which is now called complex partial epilepsy. Complex because the activity that the epileptic discharge produces is of a complex nature and partial because only a portion of the brain is participating in the abnormal activity.

Her original neurologist wondered if she might have a brain tumor. He had been worried enough over this possibility that Lucy and her mother had translated his concern into a probability and then an almost certainty. (Fear does strange things to perceptions, which doctors remain aware of at all times. Patients may not hear what doctors are really saying. Often doctors blithely give out a long list of possible diagnoses, some unlikely possibilities to be considered and then discarded once the correct diagnosis is made.) A brain tumor in this case was certainly a possibility, but it was not the most probable diagnosis.

Traditionally, we neurologists have been a strange breed. One of the oddest things about those mavericks-turned-neurologists who created our field is that they developed a strange view of the world of the brain, and along with it a unique logic for constructing and analyzing that world. Those of us who love that peculiar paradigm become neurologists.

Much of this thinking process was necessary because there was no way for nineteenth- or early twentieth-century neuroscientists or clinicians to study the workings of the intact brain. There were no PET scans to show which areas of the normal brain was affected during a particular function and therefore must be playing a role in the physiology of that function. Today we can do that. Less than two decades ago we couldn't. The most remarkable part is that, now that we can study normal brains, we have done little more than confirm what neurologists already figured out by studying patients whose brains were obviously not normal at all.

Most of what we neurologists know about the brain and use daily to make diagnoses in our patients is still based entirely on the study of brains that didn't work right in some way or other.

This is not surprising since neurologists are asked to see patients only when something has gone wrong. We examine such patients much as neurologists have examined patients for the last hundred years or more, and we perform that examination in order to define what has gone wrong within the brain as carefully and completely as we can. We try to answer a single question: Precisely what function or functions have been lost?

Once this has been determined, we use that knowledge to answer a second question: Where within the brain has something gone wrong? Put another way, what is the precise location of the loss of function? What is the anatomical basis of the altered physiology? Initially this question was answered anatomically by studying the brains of previously examined patients after the patients had died, a process known as "brain cutting." This is how Paul Broca, a surgeon and amateur anthropologist, localized the area involved in loss of speech. Broca had examined a patient who had lost the ability to speak. The patient died not too long after Broca had examined him, and when Broca looked at the brain of that patient without ever cutting it open, he found an area of destruction on the left side of the brain in the back part of the frontal lobe. Broca then determined that this area was responsible for the ability to speak. He did not announce his monumental discovery at a neurological meeting. There were no neurologists then and, as a direct result of this obvious deficiency, no neurological societies to hold meetings. Broca made his announcement at a meeting of the French Anthropological Society because they were interested in the ascent of man, the development of speech, and other skills specific to humankind.

These anthropologists were well aware that the two human traits most related to the evolution of modern humans were handedness, which is responsible for our ability to make tools and perform other tasks leading to success in our struggle to survive, and speech, which is the most important attribute of the human nervous system. Both of these traits are based on the structure and function of our brains and are the stuff that neurology is made of. And the area that Broca described to those anthropologists even to this day is called Broca's area.

The modern clinical examination is not very different from what physicians did in Broca's day. But we no longer have to make our best guess as to the localization of the lesion within the brain. Now we see images of the brains of living patients with CT scans and MRI scans. You can always tell the age of a neurologist who is discussing a patient with you over the telephone by observing which he mentions first: the examination or the CT scan. If the scan come first, he's under the age of forty. If the CT scan comes after the neurological exam, he's over forty and was trained to think like a classical neurologist. The process of going from neurological examination to identifying which part of the brain is injured or diseased is called cerebral localization.

But the process of neurological reasoning goes one step further. If a specific function is lost when a particular area is injured, then that area when not injured must be the site at which that function is performed. (For example, Broca's area is not merely the area which, when injured, causes a loss of speech; it is also the area that, when the brain is intact, controls normal speech.) Cerebral localization, begun as a way of figuring out which part of the brain is involved in a disease process, became the basis for mapping the functions of the normal brain to its parts. This theoretical framework constituted a great leap forward in both faith and logic, a leap that attached normal function to the area that was injured when that function was lost.

Neurologists were not the first to presume that different parts of the brain control specific functions. That is, the brain's not like the liver. True, the liver has some things in common with the brain. It too is a collection of cells which may vary in size but look pretty much like each other no matter where you are in the organ. One look at a slice of liver tissue under the microscope, and you know that you are in the liver. So too the brain. But all liver cells do pretty much the same thing. Not so when it comes to brain cells. Why don't they work as a whole and not as subunits? How simple neurology would be if the entire brain did whatever it is the brain does, not each area doing its own thing.

It was the phrenologists who first emphasized the concept of the localization of brain function. That's right, phrenologists—those humbugs who claimed they could analyze brain functions by palpating the bumps on a patient's head. One of the smarter phrenologists had studied the inner surfaces of the skulls of human and various other primates and had decided that speech was localized in the back part of the left frontal lobe—precisely where Broca found the lesion in the brain of his patient. And Broca was looking for the lesion there because he was well aware of this theory.

But the *way* the phrenologists' defined the localization of cerebral function is where phrenology and neurology parted company. The logic of neurology is the logic of Paul Broca: find the area in the brain that is not functioning, and you have located the center of that function when the brain does work. Peculiar as that logic may be, it remains the best view in town.

This was where we stood when Lucy Arft entered my life. Unintentionally this young woman would introduce me to the entire notion of windows of opportunity as well as to the neurology of handedness and speech.

It was clear to me that Lucy had epilepsy. Neurologists use the term "epilepsy" to mean simply a recurring tendency to have seizures. It has no other specific connotations. Lucy had dozens of seizures before she had seen any doctor. Sometimes she had eight or ten a day. Rarely did a day go by without her having at least one, despite the fact that she was on large doses of anticonvulsant medications. Seizures that cannot be controlled are often the sign of a serious underlying problem, such as a brain tumor. No wonder everyone was so damned worried.

What then is a seizure?

A seizure is an event. A change in behavior. An episode of abnormal behavior, such as smacking of the lips and becoming unresponsive, that is the direct result of excessive electrical activity of the brain. In some seizures, the entire brain seems to be firing at once and the abnormal discharges cover the entire cerebral cortex. That is the classic generalized convulsion, the true grand mal seizure. In other cases, only a part of the cerebral cortex is abnormally firing away. In these cases, there is only a localized area of abnormal electrical activity and a far more limited form of abnormal behavior. It is still a seizure but not grand mal. It's a partial seizure, or in neurological parlance, a focal or local seizure.

If a patient has seizures that start with abnormal jerks of the left leg, then the seizure focus is in the leg area of the right frontal motor cortex. If the patient calls out and then stops speaking altogether, then the seizure focus is probably in Broca's area, in the frontal lobs. The phrenologist might feel there. As a neurologist I would look there. This is our old friend cerebral localization at work.

The type of seizure Lucy had is also related to a local form of abnormal electrical discharge in this area that receives and interprets odors. Such seizures emanate from the temporal lobes. Hence the abnormal discharge produces a smell. (An abnormal discharge in the visual cortex would produce a visual image, and so on for the other senses.)

Abnormal discharges starting in the temporal lobes indicate the presence of lesions on the temporal lobe. Ergo, Lucy Arft, this nineteen-year-old University of Michigan honor student, had something amiss in one of her temporal lobes. It was my job to figure out what type of lesion Lucy had and in which temporal lobe. (Both lobes cause identical types of seizures.)

As I took her history, I observed her carefully. By the time I had heard her entire story I knew she probably did not have a brain tumor. What she did have was far more interesting than the uncontrolled growth of abnormal cells gone berserk. For Lucy Arft, my very own Lucy, was the only left-hander in her entire family. Both of her parents, all four of her grandparents, her two brothers and three sisters, and every first cousin were right-handed. One hundred percent. So Lucy should have been right-handed too. So what? you might ask. Actually, this made her the stuff of neurological history.

Handedness has nothing to do with our hands. It has to do

with our brains. You cannot look at a patient's hands and tell whether that patient is left-handed or right-handed unless there are some work- or habit-related changes on the skin that leave their telltale marks, such as ink on the right hand of a scrivener or smoke residue on the fingers of a left-handed smoker. It is the motor skills of the two hands that are different, not their strength or their structure, nor the ability of the hand to carry out fine, rapidly alternating movements or other highly coordinated activities. Degree of agility is what makes the difference.

Anatomically, hands are the perfect example of mirror-image structures. They are in fact the model for classifying enantiomers in organic chemistry. Enantiomers are a particular kind of isomers, organic substances that come in different structural forms but have identical compound formulas. Two enantiomers bear a particular geometric relationship: one is right-handed while the other is left-handed—they are mirror images of each other. This handedness is possible because their formulas include carbon; the carbon atom is symmetrical so the atoms attached to it can be on either side. Enantiomers have different characteristics, and thus produce a pair of identical isomers with far different functions.

How different? Most of the compounds we use in our bodies are left-handed isomers, while their right-handed partners are inactive. Consider, for example, dihydroxyphenylalanine, or dopa. The left-handed enantiomer, levodopa, is the single most effective treatment we have for Parkinson's disease. When levodopa enters the brain, it is converted into dopamine, the chemical in short supply in the brains of patients with Parkinson's disease. This newly formed dopamine improves the patient's physical disabilities, which were due to the brain's inability to manufacture enough dopamine on its own. Levodopa's complementary isomer, dextrodopa (right-handed dopa), is inert: the enzymes of the brain cannot attach to its active site, so it cannot be converted to dopamine; it does the patient no good at all.

Like the chemical structures of enantiomers, the two hands of a patient look identical, or as identical as mirror images can be.

But once the patient is examined and the motor skills tested, the differences between those outwardly identical hands becomes obvious. At times even the explanation for those differences becomes apparent.

But it is misleading to say we are right- or left-handed. The basis of our dexterity lies not in our hands but within our brain, which controls the movements of our hands. This means that, as far as manual dexterity is concerned, we are either left-brained (and thereby right-handed) or the opposite. According to the way neurologists view the world, handedness is just another of the brain's acquired nonverbal motor skills. Although handedness is basically inherited, we are not born handed; we acquire handedness. This statement is not as contradictory as it seems, as will become evident shortly.

About 90 percent of the population is right-handed. Those who are right-handed not only prefer to use the right hand but are far more skillful in using that hand for a wide variety of learned skilled activities, including writing, drawing, and (in the United States) throwing a baseball. This difference in function is not paralleled by any difference in the structure of the two hands. Thus, unlike enantiomers, handedness has no structural basis but is dependent entirely on brain *function*.

Such right-handed preference is universal; it is characteristic of the entire family of humans. That includes all cultures from the Stone Age to the information age, regardless of language group, the way a language is read (right to left, left to right, or even top to bottom), or any so-called racial characteristics. Given that universal preference, why do some 10 to 12 percent of the population come up left-handed, generation after generation, all over the world? Some left-handers, of course, become left-handed because something goes wrong in their left hemisphere which prevents them from becoming right-handed. Neurologists use the term "pathological left-handers" to define this situation. I recognized that Lucy was left-handed almost as soon as I met her. She wore her watch on her right wrist. This then posed a basic question: Was her handedness hereditary or pathological? Did it represent her genetic makeup, or damage to her left hemisphere? The latter is far from being the more common reason, even in patients seen in the offices of those of us who spend our lives studying diseased or damaged brains.

In most left-handers the fact that the individual becomes right-brained for handedness is not the result of brain injury but is genetic. The most attractive genetic theory was proposed by British psychologist Marian Arnett in 1985, who suggested that genetic variations in handedness could be due to a single gene that can exist in either of two alternates (or alleles) which usually arise through mutation. The dominant allele produces a "right shift" in those who possess it; that is, it increases the likelihood that they will become right-handed. This dominance accounts for the fact that the distribution of handedness is heavily in the direction of right-handedness. Some very small fraction of this population would still become left-handed due to environmental influences, but that number is not large enough to explain the 10 percent incidence of left-handedness in all human populations.

That is where the other allele comes in. The recessive allele does not cause left-handedness *per se* but removes any *bias* toward handedness, so either side can become dominant. This explains why the children of two left-handed parents are themselves equally divided into left- and right-handers, and also explains why left-handers show a mixed pattern of asymmetry on other measures, such as eye dominance, footedness, and even fingerprints.

Since her parents were both right-handed, Lucy would have been far more likely to have inherited the allele for a "right shift," or in neurological terms a "left brain shift for dexterity."

Yet she was left-handed.

Was she therefore a pathological left-hander? And if so, when had this pathological condition started?

Handedness is acquired quite early in life, and like language acquisition demonstrates another window of opportunity. An adult who has a stroke resulting in severe paralysis of the dominant right hand can learn to use the left hand for skills previously performed with the right hand. He or she can learn to write, use a fork, and perform a wide variety of tasks. However, these tasks are never performed as skillfully as they were with the dominant right hand. Children, however, can switch dominance quite easily.

That in fact is the neurological basis for the concept of a gene that merely causes a shift or a change in the odds. At birth either hemisphere can become dominant for handedness. Normally, because of genetic dominance, it is the left hemisphere. But if something goes wrong in the left hemisphere, the right hemisphere will do, and it will do just as well as its mirror image.

Although the studies haven't been as numerous as those for speech, the rules for the recovery from hemiplegia (including switching of dominance for handedness) parallel the rules that apply to reacquiring speech. The brain, on the basis of genetic inheritance and environmental input, selects a dominant hemisphere for handedness. If this hemisphere is injured early in life, the brain can then select the opposite hemisphere and switch dominance. After puberty this becomes increasingly difficult, and in adults it cannot be done. Abilities can be acquired by the noninjured hand and the nondominant hemisphere, but true skilled handedness can never be acquired by what the person had grown up with as the nondominant hemisphere.

Pathological left-handedness occurs when an individual becomes left-handed because brain injury to the left hemisphere early in life prevented the development of the (hereditarily dominant) right hand. Did Lucy have a lesion of the left hemisphere that made her into a pathological left-hander? And what had happened to her speech? In most of us the left hemisphere is also dominant for speech. Broca's area. Even the phrenologists knew that. But her speech and overall intelligence were obviously normal. She should have had a dominant left hemisphere. If she didn't, why not? And when had the switch in dominance occurred?

My neurological examination of Lucy revealed that her right hand was smaller than her left hand: the nail beds of her fingers were narrower, and her digits were shorter. Her hands were not true mirror images of one another. No racemic isomers here. I saw the same asymmetries in her feet. Her right foot was smaller, with narrower nail beds and shorter toes. Eureka! I knew exactly what was wrong with Lucy's brain. These asymmetries were the classic signs of hemiatrophy, or the failure of one side of the body to reach full growth. They are only seen when the brain injury or disease occurred before the age of puberty— almost always several years before puberty, when the body is still growing, and long before the end of the closure of the window of opportunity for handedness and the analogous period for speech. But when had it occurred to Lucy, and what had happened to her brain? Lucy could remember nothing. As far as she knew, she was normal and always had been. Her academic accomplishments were proof that she had never suffered any brain damage whatsoever.

Like all mothers everywhere, Mrs. Arft knew far more than anyone about what had happened to her daughter early on. So we started at the very beginning. The pregnancy had been normal. The labor and delivery had not been. Lucy had been a breech presentation. After a prolonged labor lasting over twenty-four hours, Lucy was finally delivered by an emergency cesarean section and had remained in an incubator for almost a month. Shades of Julius Caesar himself. The very term "cesarean section" derives from the emperor. His mother died during a prolonged labor, and he was delivered through surgical intervention. Later in Caesar's life he, like Lucy, developed epilepsy, usually attributed to brain injury which occurred at the time of his birth. Lucy most likely had suffered the same sort of insult to her brain.

I continued to pursue Lucy's history. As an infant Lucy dragged her right leg when she first walked. This was thought to be an orthopedic problem, and orthotics (special shoes) were prescribed for her. And even after she no longer required the orthotics, it had still been difficult to buy her shoes. Her right foot was always a size or two smaller. Often two pairs in different sizes had to be purchased to get a pair that fit.

This history made it clear that Lucy's right-sided hemiatrophy dated back to birth. Like Caesar's birth injury, the injury to Lucy's left hemisphere was most likely the result of an abnormal deprivation of oxygen during the late stages of pregnancy and delivery. Since the ability of her left hemisphere to control her right arm and leg were abnormal, Lucy became left-handed. She was indeed a pathological left-hander.

The logic was impeccable. Early injury to the left hemisphere resulted in right hemiatrophy and seizures. But were her seizures coming from that injured left hemisphere? Complex partial seizures are identical no matter which hemisphere they originate in. And having an old injury in one hemisphere does not prevent new disease from starting in the other previously normal hemisphere. Even the logic of cerebral localization has its limits. Lucy might have a tumor on the other side of her brain.

That was a question that could be answered, even in the antideluvian era before CT and MRI scans. We did have the ability to do EEGs, or electroencephalograms. An EEG is a record of the electrical activity of the brain. Lucy had already had several EEGs, and a study of them revealed abnormal bursts of electrical activity, all of which emanated from her left temporal lobe. Cerebral localization still worked.

I told this to both Lucy and her mother, who both understood exactly what I was saying. Her mother began to cry with relief and of joy: Lucy did not have a brain tumor. Thank God, no one would have to operate on her little girl's brain.

But could I make her seizures go away?

I didn't think I could, at least not with medications. She had been treated with large doses of every available anticonvulsant, and she was still having eight to ten seizures every day.

Wasn't there anything else I could do?

There was.

What?

Brain surgery.

Brain surgery! It was as if I had pulled the rug out from underneath both of them. They were crushed. And confused. Lucy couldn't need brain surgery if she didn't have a brain tumor. But she did need brain surgery; she had seizures.

I took a deep breath and tried to explain it to them as gently as I could. The scarring of her abnormal left hemisphere was causing the damaged cells there to produce seizures. If medications couldn't control those discharges, we could take out the scarring and those abnormal cells.

How dangerous was that?

That all depended on which hemisphere controlled her speech, which side of her brain was dominant for speech. We went through it again: Like handedness, speech is usually controlled by the left hemisphere. Children who suffer major injuries to the left hemisphere early in life lose the ability to speak, but they recover fairly quickly. It is not that their left hemisphere repairs itself. Their brains switch dominance, so that the ability to speak becomes controlled by the uninjured right hemisphere.

Just like handedness.

And like handedness this capacity to switch dominance is governed by age. By puberty it has waned; the window of opportunity is closed.

Lucy's injury occurred long before puberty. She may never have switched dominance. But which side was dominant? If it was her right hemisphere, tapped for speech because it was needed before the window of opportunity for speech closed, then operating on her left hemisphere would carry no threat to her speech. If her left hemisphere, although injured, was still in control of her speech, surgery there might be too dangerous.

How could we tell?

By performing a Wada test. This procedure is named after Juhn A. Wada, the Canadian neurologist who perfected it. The test consists of injecting a small amount of a rapidly acting barbiturate into each carotid artery and then observing its effect on speech. Each carotid artery supplies blood to only one hemisphere, the left carotid to the left hemisphere, the usual location of Broca's area, and the right carotid to the right hemisphere, the location of the usually inactive mirror image of Broca's area. When the activity of the dominant Broca's area is arrested by the barbiturate, speech stops. When the drug hits the other side, nothing happens.

Three days later we performed the Wada test on Lucy while monitoring her EEG. The injection into her left carotid artery put the left side of her brain to sleep but had no effect at all on her speech. Lucy continued to talk to us as we watched the electrical activity wane on that side of her brain until all that remained were the abnormal discharges. We had put her brain into a state in which all normal activity stopped and the seizure discharge just kept on firing. Like that damn rabbit on the Energizer battery commercials. In Lucy, Broca's area was not in her left hemisphere. That was a good sign.

A subsequent injection into the right carotid caused severe aphasia. Since only the right hemisphere was put to sleep by the injection, Lucy remained both awake and alert but she could not comprehend even simple instructions and could not speak at all. Eureka! We had found her Broca's area. On the "wrong" side. Just like her hand dominance.

As we have already noted, genetically speaking, Lucy should have been right-handed, and her left hemisphere should have developed normal speech capability. Instead her right hemisphere developed these capabilities. During her entire critical period, the normal right hemisphere was selected to develop speech. And it responded.

The rest of her story is equally dramatic and even more rewarding. The next week she had the brain surgery she had come to me to avoid. Not exactly the same surgery. A small scar was removed from her damaged left hemisphere along with the surrounding areas of her left hemisphere that were producing the seizure discharges.

As soon as she woke up, we tested her speech. It was as normal as ever. Nothing had changed. Except for one thing: she never had another seizure.

I often think of this Lucy of mine, in part because she still sends me cards every Christmas, but mostly because of what she taught me. I had studied all the theories of brain development and cerebral localization long before we ever met, but she made them all come to life for me. Now I truly understood them and their implications. Early in our lives our two hemispheres are identical. Like the original Lucy, we are capable of bipedal walking, of standing upright on our own two feet. As it did in our evolution as a species, bipedalism comes first in our development. Bipedalism with two equal cerebral hemispheres, and hands as well as legs that are enantiomers in both structure and function. Then, as the hands become used for more and more activities unrelated to mere mobility, the process of selection begins, selection for dominance for handedness. This happened during the life of Lucy Arft as it happened during our evolution as humans.

Why did upright posture and bipedalism triumph in this one unusual evolutionary line? The conventional story involves a changing environment and the need for early hominids to survive in the African savanna. According to this version of evolution, it was the shift from life in the forest to life in a more open habitat that forced our predecessors to adopt bipedalism since the upright posture allowed them to see farther over the tall savanna grasses. Or perhaps to escape predators better. Or walk more efficiently over greater distances. Or perhaps the upright posture left less of the body surface exposed to the heat of the sun.

The problem is that not all examples of *Australopithecus afarensis* came from the savanna. Lucy came from Hadar in Ethiopia, a dry savanna, but her cousins who left those famous footprints as well as parts of numerous skeletons came from Laetoli in Tanzania, an area of rivers and woodlands. So why did we—or rather, why did our ancestors—become bipedal? What made this different adaptation succeed for the one line of species which developed increasingly large and complex brains?

As a neurologist I would argue that the two are not unrelated. And that the development of bipedalism did not set the stage for enlargement of the brain any more than enlargement of the brain set the stage for bipedalism. These two developments made a system of positive feedbacks, each process favoring the other but with bipedalism leading the way.

Why? Because bipedalism freed up the arms and hands. Not for making tools. That only came eons later. For what then? The answer seems obvious to a neurologist. We are unique among species in the degree of juvenilization of our brains. This is probably based in part on the size limitations on the fetal brain as determined by the dimensions of the human pelvis. Juvenilization is what has made the human brain what it is today. But that same process has made the human infant more dependent and for longer than the offspring of any other species. So how could these dependent offspring be protected?

By freeing up mothers' arms to hold them, to protect them, to keep them out of harm's way far more effectively than any other available method. And for far longer. A frightened infant chimp can hold on to its running mother. A human mother can easily pick up and carry her five-year-old child, who might not yet be smart enough to know that it should be holding on for dear life.

So first comes bipedalism, then the development of the juvenilized and protected human brain. One element of this process has resulted in the left brain shift in dominance for handedness. Then comes the development of speech. Obviously both sides of the brain can't control speech at the same time, so the same sort of shift has to take place, which involves selecting one of two hemispheres with equal potential. But this selection is not final and irreversible. It can be changed—but only for a while, until puberty.

The window closes for a reason, I believe. Once the developed brain is theoretically "on its own," no longer being swept out of danger by its mother. This closure evolved not as a result of simple cause and effect but as a process. By the time we needed to survive without mom's protection, our brain's systems for interacting with the world had to be set in place. If not, then when in danger we could not have duplicated what our mothers did and thus protected ourselves. So having the window stay open beyond puberty wouldn't have served any purpose. We wouldn't have lived long enough to use it.

How does this all happen? How do windows close? What

happens to the juvenile brain as it enters adolescence and adulthood that makes it less plastic?

I learned the answers to these questions long after my Lucy had gone on to become an associate professor of romance languages and a mother of three daughters—all of whom are right-handed.

3

—

THE GIFT OF SPEECH

.
.

Frank Morrell and the Treatment of
Acquired Epileptic Aphasia

MARILOU ATLAN WAS never actually my patient, but few of my patients, even those I have seen and examined dozens of times, have occupied my mind more than she. I first met Marilou when she accompanied her mother and grandfather to my office for treatment of his Parkinson's disease. Just over three years old, she was clearly a precocious child, with a strong will and mind of her own. Before I even had a chance to introduce myself, she explained to me that she was named after her late grandmother Mary, whom she had never met, and her grandfather, Lou. Marilou then explained in a rush that she had come with her grandfather to make sure that I would not do anything bad to him—she had already lost one grandparent and had no intention of losing another one. Here she was going on forty. Having said her piece, she sat down and allowed me to my turn.

My bit turned out to be relatively straightforward. Her grandfather, Louis Fournier, had recently been diagnosed as having Parkinson's disease, and he and his family had decided to come to see me. My examination of him confirmed he was indeed in the early stages of Parkinson's disease. I told him there were no tests to do; Parkinson's disease is a clinical diagnosis, meaning it's diagnosed through observation.

He already knew that much.

Then why had he come to see me?

"I have a Mercedes," he began. "It cost me a lot of money. When it needs service, I can go to the guys at the corner. I've known them for years. They never charge me very much for what they do. Or I can take my car to the dealer. He's half an hour away and much more expensive. What do I do? I go to the dealer. They know what is best for my Mercedes. You know Parkinson's disease. So I am here."

I understood. I see seventy or more patients with Parkinson's disease every week.

"Besides, I can always buy a new car."

We talked for a few more minutes about what he could expect from his disease and how often he should come back to see me. "Every three months," I told him. His job was to bring his Mercedes in regularly for routine maintenance.

He smiled. "I never miss scheduled maintenance."

"Grampa was scared to come here today," Marilou piped up. "But I told him he was being silly."

"Don't you like it when your Grampa acts silly?"

"Not that kind of silly, silly," she informed me.

With that said, I showed her how I could pull my thumb into two parts. She watched carefully but made no real response. Then after careful consideration, she said, "That is a very good trick. It would scare a lot of my friends. They are real sillies. Will you teach me how to do that?"

I did. When I was done, I reminded Mr. Fournier to schedule an appointment in three months.

"On a day when I can come," Marilou said.

FOR THE NEXT year, I saw Louis Fournier, his daughter Terri, and his granddaughter Marilou every three months. By the time she was four, Marilou had mastered my entire repertoire of little tricks that had taken me a lifetime to learn and, of course, she was already reading. It was a skill she had taught herself with just a little help from her grandfather.

Unfortunately Louis was not doing quite as well. His disease was progressing far more quickly than it did in most patients. We know that different patients progress at different rates, but we don't know why. He was shuffling his feet, and his balance was less certain than he liked it. I started adjusting his medications and seeing him once every two months.

By the time Marilou was five and reading everything she could get her hands on, Louis Fournier was losing his balance and falling at least once a week. I thought the best alternative was to place him on an experimental medication. While I was discussing it with him, Marilou seemed not to be herself. Normally she sat quietly, not moving a muscle, deeply involved in making sure she heard every word I said so she could repeat my words to her mother and grandfather if either had any problem recalling them.

On this particular day, she didn't seem to be paying as much attention as usual. I wondered if this was because she was growing up and becoming more concerned with her own life. Suddenly, out of the corner of my eye, I noticed that Marilou blinked her eyes rapidly several times, then jerked them briefly to the left and dropped her head for a second. In the next moment, she rubbed her hands together quickly. By the time I turned to look at her, she gave me her usual smile indicating everything was fine.

As she took her grandfather's hand to lead him off to make his next appointment, I asked her mother if Marilou was okay.

"She's been strangely quiet lately. It's sort of a relief. We're all worried about Dad, especially Marilou. If he has a bad day, she is so worried about him, and he's had so many bad days recently. Plus she's in a new school and having trouble making friends. She is more advanced than everyone her age, but she's too young to be pushed ahead. I think that's why she's been quiet."

Then Marilou looked at me and asked, "When?" It was the first time she'd spoken during the whole visit.

Obviously she had been lost in her own thoughts and hadn't heard me earlier tell her grandfather to come back in a month. "Two weeks," I said this time.

"I thought you said a month," her mother said.

"I changed my mind." I was worried about Louis Fournier, but I was also worried about Marilou. What was happening to the vibrant young girl I had come to care about as if she were my own daughter?

TWO WEEKS LATER, Louis Fournier was showing signs of improvement after taking his new medication. He no longer shuffled his feet when he walked, and his balance had improved. I made some further adjustments in his new medication, increasing it from three to four doses each day. Normally I would tell such a patient to come back in six or eight weeks. "Two weeks," I said.

Then I turned my attention to Marilou. She had not yet uttered a single word. Hard as I tried, I could not get her to say anything to me. I did all my old tricks. She watched them and smiled but said nothing. Not one word.

Once again she took her grandfather by the hand and walked him out to the front desk.

"Terri," I began.

"I know. She's going to see a psychiatrist next week."

"A psychiatrist?"

"She seems so depressed. Terri mentioned the child psychiatrist their pediatrician had suggested.

I approved. The psychiatrist, Dr. Rolland Pignon, had an excellent reputation. I had sent several of my patients to see him. "What's been going on?"

"She's in her own world. At first she seemed as if she just didn't want to listen to us or talk to us. Now she won't talk at all. She doesn't seem to care what we say. She doesn't even pay much attention to her dog. She loves that dog—or at least she used to. Now, I don't know." Terri started to cry.

There were so many questions I wanted to ask, but I didn't want to alarm Terri. "When are you seeing Dr. Pignon?"

"On Monday afternoon."

"Then I want to see your father on Tuesday morning. We'll set up a long appointment."

"Thank you," she said. "We'll all be here."

The next Tuesday morning, the three of them arrived at my office at eight-thirty. The new protocol had made all the difference in the world for Louis Fournier. He was almost back to his old self.

But Marilou was a shell of her former self.

Dr. Pignon was certain that she was depressed, Terri told me, but he had some other concerns and wanted Marilou to see a pediatric neurologist. Since the Atlans already had an appointment with me, Terri hoped I would take a look at Marilou.

"The first thing I noticed," her mother said, "was that she didn't seem to pay as much attention to me. Sometimes it was as if she didn't even hear me."

"As if she didn't understand you?"

"That too, but I'd say something, and it was like she never even heard it."

"Just with words? Does she pay attention to other sounds?"

Her mother didn't know but her grandfather did. "She still plays with Marty. That's her dog. 'Marty' is short for Charles Martel. But if Marty is in the next room and barks, Marilou doesn't even look up."

Now I was worried.

"When did it start?"

"Six weeks ago, when the new school year began."

All in six weeks. From normal speech to a total loss of all speech functions, both receptive and expressive: she could not understand language, and she could not produce it. She was mute.

I wondered if Marilou could hear, so I shouted at her.

She said nothing.

I walked behind her and dropped a heavy book on the floor.

Her mother was startled. So was her grandfather. Marilou sat there as if she hadn't even heard the sound.

I stood in front of her and did my old trick, pulling my thumb apart. Marilou smiled, and as usual did all of the tricks I had taught her. She even showed me a couple her grandfather had taught her. Marilou then took a deck of cards out of her pocket and showed me two card tricks.

"When did she learn those?" I asked.

"Last week," her grandfather informed me.

I was dumbfounded. Marilou couldn't understand a word anyone said to her. She failed to even recognize the concept of verbal communication. She didn't respond to sounds, not even to the bark of Charles Martel. Yet she could learn complicated card tricks, ones that depended on interactions between people that required her to learn complex, sequential behaviors. Obviously most of her brain was still functioning quite well.

As I tried to figure out exactly what was going on, Marilou did it again: in the same sequence of motor behaviors, her eyes blinked rapidly and jerked to the left; her head dropped, and she rubbed her hands together. I knew exactly what I was observing: a seizure.

Her eyes jerked first to the left, then back toward the center, over and over again: left, center; left, center; left, center—all the while her eyes kept on fluttering.

Then her eyes stopped fluttering. At almost the same instant, they stopped jerking and floated back into their normal position.

But it wasn't over.

She put her two hands together and began to move them as if she were washing them. Over and over again. I timed this: she continued washing her hands for about forty seconds. Then she stopped. The entire seizure had lasted no more than a minute. It had started with eye movements and then had concluded with what is called an automatism, a form of automatic behavior. The term "automatism" is used to describe a complex movement or set of movements that would be considered normal done at the right time and in the right place but is instead performed at the wrong time or in the wrong place and serves no apparent function—such as rubbing one's hands together when it isn't cold. Such automatisms are usually seen in seizures that emanate from the temporal lobe.

Marilou's failure to understand words indicated she might have aphasia, or loss of the use of language. In particular, she seemed to be exhibiting signs of receptive aphasia, meaning her ability to understand language was more affected than her ability to produce it. This condition pointed to a problem with her dominant hemisphere for speech, and since she was right-handed, the problem would be located in her left hemisphere. Like a real estate agent, a neurologist has three primary concerns: location, location, location.

But where in the left hemisphere? The left hemisphere is a big place. It's one half of the brain. Receptive aphasia is usually found more toward the back of the brain than the front, in a region known as Wernicke's area. This region is named after Carl Wernicke, the neurologist who first described patients with receptive aphasias and who demonstrated that this condition is localized in the posterior region of the left temporal lobe. A Wernicke's aphasia due to disease in Wernicke's area, a circle of eponyms.

What causes aphasias? Usually strokes. But sometimes tumors, aggressive, malignant brain tumors.

However, Marilou did not appear to have a classic receptive, or Wernicke's, aphasia. Such patients usually have the same sort of trouble understanding language that her mother described to me. But they still pay attention to anyone who speaks to them. They still retain the concept that language has to do with communication and that when someone speaks to them, that person is trying to communicate with them. They listen and they respond. It's just that they have no idea what has been said to them, and their answers also make no sense. Patients ramble on and on saying lots of words, too many words. And Marylou was silent.

If she wasn't suffering from classic Wernicke's aphasia, then what had brought on the seizures and caused the aphasia?

What was going on with this beautiful little five year old?

I still felt certain that Marilou's seizures came from the cerebral cortex of her Wernicke's area. (The cerebral cortex is the outer layers of gray matter of the two hemispheres.) She had a seizure focus, or origin, there, right inside her center for receptive speech.

"Her birth was normal?" I asked.

"Absolutely."

"And your pregnancy?"

"Yes, yes. She was such a beautiful child. What—"

"And she met all of her developmental milestones," I surmised. "In fact, she probably started talking very early."

This brought a pair of nods.

"For five years she was perfectly normal?"

"She was not just normal," her grandfather said. "You saw her and spoke with her, listened to her talk. You know how bright she was."

Then at five, her problems had started. This part of the story I knew. At first she had difficulty understanding words. It was as if no one could get through to her. Perhaps she was tired, or wasn't paying attention, or coming down with something. But within a week or so, the problem was clearly more than that: now she not only couldn't understand words, she couldn't produce them correctly, used the wrong words. She began speaking less and less until she stopped speaking altogether. At this point she no longer responded to words; it was as if they were just other sounds to her. Finally she stopped responding even to other sounds, to Charles Martel's bark.

"If only I could hear her talk again . . ." her grandfather said.

There were only two dry eyes in the room and they both belonged to the one person we were all talking about. She sat there as if language had no meaning to her. As if she were Genie or the Wild Boy of Aveyron, and someone had shut her away in some dark closet. In a way that's exactly what had happened. Her own brain had shut her away from speech.

"She has what we call Landau-Klefner syndrome. Or at least that's what I think she has."

"She has what?" her mother asked.

I was not an expert on Landau-Klefner syndrome. I had never made the diagnosis before, and technically speaking I wasn't making it now since Marilou was not formally my patient. But I did know someone who could help her, the neurologist Frank Morrell, an old friend who specialized in epilepsy just as I specialize in movements disorders. He also was an expert on the cerebral cortex. Those two subjects go hand in hand since most epileptic seizures are caused by abnormal electrical discharges of cells in the cerebral cortex.

Landau-Klefner syndrome is the perfect example of having the wrong problem in the wrong place, and frequently at the wrong time in life. The wrong disease is focal epilepsy, seizures that start out as sudden electrical discharges of a group of abnormally firing cells of the cerebral cortex. The wrong place is the speech cortex. The wrong time is during the window of opportunity for acquiring speech. The result is a total disruption in language acquisition. Worse than that, it's a total loss of language, a state of global aphasia, in a disease sometimes called acquired epileptic aphasia. "Acquired" means that this syndrome happens to a previously normal child: the patient was once able to speak, to use language, but then lost that skill.

By now I was thinking out loud, trying to tell Terri and her father what I understood to be happening inside Marilou's brain. Or more correctly on the surface of Marilou's brain.

Landau-Klefner syndrome starts early in life, always after speech has started to evolve—after the age of two—but before the age of six, before language has been completely established.

Marilou was only five. She was right in the middle of the critical period for speech acquisition, when her cerebral cortex was reinforcing all of the connections her brain would use for the rest of her life for both understanding and speaking language. As Lacey's story told us, this reinforcement process must be put into place before this window of opportunity closes.

"When is that?" her grandfather asked.

"About age twelve."

"We must hurry."

I smiled. "We have time. I think she'll be able to speak again.
The man who can help her is right down the hall."

These were the first words of real hope that I had given to
them. It was as if the entire Magnificent Seven had arrived all
at once to save the day, along with Harrison Ford, Clint
Eastwood, and Arnold Schwarzenegger.

Her mother was still dabbing her tears away.

Marilou was drawing a picture of her dog, Charles Martel, a
dog who couldn't bark. It was the kind of drawing you might
expect from a precocious eight year old—not from a five year
old who could no longer utter a single word.

"Patients with Landau-Klefner syndrome also have seizures,"
I continued. "In most patients they are pretty much the same."
I then described a typical seizure. Eye blinking. Deviation of the
eyes. Deviation with jerking eye movements. A dropping of the
head. Then some sort of automatism.

Had Marilou ever had a real convulsion? A generalized
seizure?

They both insisted that she hadn't.

Meanwhile, Marilou continued to draw. It wasn't as if we
weren't there. Whenever she finished a drawing, she would get
up and show it to her grandfather and wait for his smile of
approval, which she always got and which evoked a return
smile of pleasure from her. Then she showed it to her mother,
and then to me, again exchanging smiles. Always with the same
appropriate responses.

Then another sheet of paper and another drawing. Another
drawing of Charles Martel. Not a repeat. Charles Martel chew-
ing on a bone. Charles Martel playing with a ball. Charles
Martel sniffing the base of a tree. Charles Martel *ad nauseam*.
No talking.

"The seizures," I went on, "are usually easy to control. They
respond to treatment with anticonvulsants, and by the time the
patients are in their early teens, the seizures seem to go away
and the patients don't even need to be on anticonvulsants anymore."

"But will she talk again? Will she understand us?" It was her grandfather who did all of the talking now. His voice was stronger than I had ever heard it before, with no trace of his parkinsonism in it. That did not surprise me. The emotion of the moment had mobilized all of his resources and as a result his voice was that of a normal, worried, frightened grandfather.

I took a deep breath. It was time to bite the first of two bullets. Speech is a different matter from the seizures. After the latter have stopped, the speech also recovers, but only somewhat. The recovery is never complete. Rarely was it enough to allow the patient to lead a normal life. These patients always have severe lifelong aphasia.

"Alwaths . . . ?" her grandfather slurred out. The emotion had gone on too long, the effort too great, leaving him exhausted both physically and emotionally. He spoke now with the soft, slurred monotonous voice of a patient whose Parkinson's disease was taking its toll.

"You said there was something that could be done to help her?" This question came from her mother.

Now was time for the second bullet. This was where Frank Morrell came in. I tried to explain as well as I could.

Electroencephalograms usually show that the patients have seizures starting in their speech cortex. The EEGs are filled with abnormal high-voltage discharges, called spikes, that emanate from Wernicke's area on the left temporal lobe. I was certain Marilou's EEG would show this.

Some doctors have stopped the seizures and seizure discharges in a few patients by giving them high doses of steroids. And speech has returned in most of these patients after the seizures have stopped. Then the steroids are withdrawn in order to stave off unwanted side effects. But then the abnormal electrical discharges come back, the seizures recur, and aphasia once again is acquired. This means that the abnormal electrical activity and the seizures are what's causing the aphasia. Something is physiologically wrong in Wernicke's area. The cells aren't functioning correctly. They fire when they shouldn't, in ways they shouldn't be firing. It is almost as if they are too busy firing away to ever act normally. Too busy having seizures to acquire speech.

The continuous abnormal discharges had closed Marilou into the closet of aphasia and locked the door behind her.

Marilou's left temporal lobe was so busy making spikes that it couldn't do what it was supposed to do. But even though she had another temporal lobe on the other side that normally would take over, Marilou's abnormal electrical discharges also crossed over to the other side of her brain, especially during sleep, becoming even more prominent. There would be continuous abnormal discharges involving both temporal lobes. So both lobes were too busy to learn to speak.

But surgery might be able to help her. A new type of brain surgery had been designed by Frank Morrell along with surgeon Walt Whisler to stop the seizure discharges and allow the speech function to start up again.

The Atlans never asked how it was done, or why it worked. Just where should they go and whom they should see?

Dr. Frank Morrell was just down the hall. After the small family had left my office, I looked down at the series of sketches of Charles Martel that Marilou had drawn. Perhaps because he was living in a silent world, he was the saddest dog I had ever seen. Except for the last picture. In that one, Charles Martel was sleeping and smiling. She'd drawn it just as I was telling her mother and grandfather there was hope.

THE SURGICAL PROCEDURE was elegant in its simplicity. It was based on two well-established observations, neither of which originated with Frank Morrell. The first has to do with the basic functional organization of normal electrical and physiological activity in the cerebral cortex. The cortex is normally organized into vertical independent units, or columns. Each column receives its own input from below and sends down its output—up and down, in and out. Individual columns work side by side but never communicate side to side. They communicate by sending messages down and then up an adjacent column. Or to a column farther away, one that might be in another lobe, or on the other side of the brain across the corpus callosum.

That is how the normal cortex is organized—how the motor cortex controls movement and the speech cortex controls acquisition of language.

The physiological and anatomical organization of seizure discharges is not so organized. Epileptic discharges do not travel up and down vertical columns but from side to side. And there has to be a critical minimum area of side-to-side transmission in order for seizure discharges to be generated. Spikes require an area far larger than a single column, larger even than a parcel of columns. Although this understanding did not originate with Frank Morrell, he spent much of his life helping to delineate the circumstances under which such discharges occur.

The side-to-side discharges in Marilou's brain were preventing the columns of her speech cortex from carrying out the vertical in-and-out function for which they were designed. Was there a way to prevent the side-to-side discharges from occurring without disturbing the in-and-out columns?

Surgical removal of the abnormally firing neurons (nerve cells) had long been recognized as a treatment option in patients with intractable epilepsy. In so-called silent areas of the brain—those parts which do not have a specific role—abnormally firing neurons can be surgically removed with no change in function. But the cortex is not a silent part that can be cut out and sent down to the neuropathologist. Removal causes paralysis.

That was where the genius of Frank Morrell came in. His solution was to dice the functionally abnormal area of the cortex. Small, careful cuts would divide the tissue into areas too narrow to generate seizures, while leaving the vast majority of vertical columns completely uninjured. These uninjured columns would be freed of the threat of those side-to-side abnormal discharges and so could do what they were designed to do: once again in and out, up and down. A return to normalcy, as Warren G. Harding once said.

Frank Morrell and surgeon Walt Whisler mapped out and performed the surgery first in patients with seizures in their motor cortex that could not be controlled by medications.

When this was successful, they tried the same approach on adults with seizures originating in the speech cortex, usually in Broca's area. Such patients could not have conventional removal of the offending neurons without being left permanently aphasic.

Could the speech cortex be diced so as to stop the seizures without causing aphasia? Would the difference between side-to-side transmission of seizures and the vertical transmission of speech apply here as it had in the motor cortex? Or would the columns of speech tissue be so sensitive that any degree of disruption would be too much? Here the most patient of all neurosurgeons was needed—to dice where Frank instructed, and to the right depth and no deeper. Again he called on Walter Whisler, and once again it worked. Horizontal dysfunction was aborted: the seizures stopped. Vertical function was preserved: no aphasia developed. Science at work.

But would the surgery work for patients with Landau-Klefner syndrome? Clinically, their seizures were easy to control; they always went away. It wasn't the seizure that had to be prevented, but rather the underlying seizure discharges. If they could be stopped without causing any damage, then perhaps the speech would come back.

Would it work?

I watched from the sidelines as Frank evaluated Marilou Atlan. In order to prove the diagnosis, she had to have a severely abnormal EEG. She did. When she was awake, her EEG was filled with focal spikes that seemed to come from the posterior regions of both temporal lobes, although they were far more prominent on the left side than the right. In precisely the place where they should have been more prominent. When she was asleep, her EEG showed a pattern of continuous spike and wave discharges of slow-wave (CSWS) sleep in both temporal lobes, as is characteristic of Landau-Klefner syndrome.

But before Frank and his team could operate on Marilou, they had to prove that the rest of her brain was normal. The workup revealed no evidence of brain damage or dysfunction. Other than aphasia and her seizure discharges, everything else was normal. (Of course, all they would have had to do would be to watch her perform her card tricks to know that the rest of her brain was doing just fine.)

The next question was the critical one. Was there a single focus in her brain producing all of the abnormal seizures, one small area of abnormally discharging cortex that could be identified? Or several?

To find out, Frank's team of neurologists and neurophysiologists did a methohexital suppression test. Methohexital is a very rapidly acting barbiturate which causes neurons to stop firing. The purpose of the test is to identify the source of abnormal electrical discharges by suppressing all normal background activity, to produce an electrically silent brain: All normal spontaneous activity ceases; all electrical discharges dependent on vertical-column functioning stop. The background becomes a flat line on the EEG. But side-to-side epileptic discharges, which are more resistant to suppression, persist. That was what the test showed in Marilou Atlan. As her background became flat, the abnormal spikes coming from her temporal lobes seemed larger than ever. This enhancement was only apparent, however, because they were coming out of a flat background.

And they occurred on both sides of the brain. Did that mean that her discharges originated on both sides? If so, then an operation on one side would be fruitless.

More methohexital was administered in order to produce a greater level of suppression, to make the flat background activity even flatter. The discharges on the right disappeared completely. This meant that the discharges in the right hemisphere were dependent on those in the left being able to cross the corpus callosum to get to the other side. Methohexital had been able to suppress that transmission, so all of the discharges started on the left. Location, location, location. This time the right single location to make Marilou a surgical candidate.

Two weeks later she was taken to the operating room for the seven-hour procedure. Frank Morrell carefully mapped out the areas of abnormal electrical discharges on her left temporal lobe, and Walter Whisler even more carefully diced them, one millimeter at a time.

Marilou never had another seizure. The side-to-side transmission of abnormal discharges had been stopped. The day after surgery, she turned to look at the telephone in her room when it rang. She hadn't done that in two months. Before she went home, she was beginning to speak.

I continued to see her once every month or two whenever she and her mother brought her grandfather in to see me. Unlike Marilou, he did not do very well. His Parkinson's disease progressed rapidly. Nothing I did ever seemed to help. Yet each time I saw him, he thanked me as if I had given him something even more precious than his own health. The doctors had fixed his granddaughter's Mercedes. It was as good as new, and he knew it. He had brought her to the right place.

In Landau-Klefner syndrome, the cells of Wernicke's area are so busy doing the wrong thing that the right pathways are not being used. As a result they are not selected to survive into adulthood. The fact that the electrical discharges are more prominent during sleep, and involve both sides more prominently then, suggests that sleep plays a key role in the normal processes of both reinforcement and pruning. Pruning of unused synapses is done precisely to help reinforce the processes that are being used. It works just like pruning that lilac bush next to your front door. When the branches that don't seem to be doing very much are removed, the ones remaining grow better and stronger.

When the abnormal discharges were barraging Marilou's two temporal lobes, they completely disrupted all normal activity. Normal selection and reinforcement depend on normal functioning within pathways. If there is no normal speech activity, the unused neurons in the speech cortex get pruned. That was what had happened to the Wild Boy of Aveyron and to Genie. They had been deprived of speech input by their environment. Marilou had been deprived of responding to any speech input by the electrical storms within her head.

In most patients with Landau-Klefner syndrome, the seizures

eventually stop. But they stop too late, after the pruning season is over. These patients are like the Wild Boy of Aveyron: their window has already closed. Marilou was much more fortunate. She was well within that window of opportunity when Walt Whisler diced her seizure focus. She made a spectacular recovery. But did she really return to normal, what would have been normal for her? That's hard to say. But she is leading a normal life. What more do you want from normalcy?

Last year I received a letter from Marilou. She enclosed a drawing she had made for her high school art class. It is a picture of my office, drawn from where she was sitting the day I sent her to Frank Morrell. It shows me looking concerned, her mother with tears in her eyes, and her grandfather staring vigorously at me. Demanding something, insisting on something. She caught the spirit of the entire scene, of each of the participants.

How much had she understood beyond the emotions of the moment? Probably nothing.

How much had she felt? Obviously a lot.

At the bottom of the picture, she added a sketch of Charles Martel playing joyously. Her beloved dog who had been there that day in spirit only. Also at the bottom was a short note. Two words: "Thank you." And a P.S.: "Charles Martel died last week."

4

—

MANGANESE MINERS

·

Hard Wiring for Movement

PETER KRYHOSKI CAME to see me believing he had Parkinson's disease. At the time of his first visit, he was seventy-two. He believed that he'd begun to show symptoms of Parkinson's at sixty-eight, about six years after he'd retired. That was his best guess. The exact date of onset for a slow progressive disease like Parkinson's is difficult to pin down precisely. When did the decreased drive of natural aging give way to the fatigue of parkinsonism? Was the stiffness of his hands and their lack of agility attributable to arthritis or to Parkinson's disease? Few physicians can ever be absolutely certain. And even fewer patients.

"What kind of work did you do?" I asked.

"I was a welder. I welded for forty-one years. Same job. Same employer. Never missed a day's work. Not one, in all forty-one years," he answered proudly.

Salt of the earth, I mused in admiration, a real solid citizen.

"Did you work with heavy machinery in closed, poorly ventilated places?"

"Doc, we all do that," he chided me. "We all" meant welders, I assumed.

"Did you ever have carbon monoxide poisoning so bad that you passed out?" I continued, wanting to rule out carbon monoxide poisoning.

"Nope."

Exposure to carbon monoxide can cause a form of parkinsonism, but only if the victim has been exposed to levels high enough to cause a coma due to loss of oxygen supplied to the brain. If a coma has not occurred, the oxygen loss caused by the carbon monoxide poisoning is never severe enough to result in the brain damage that causes parkinsonism.

I asked him about the medications he was taking, keeping in mind that some medications can cause a Parkinson's-like state. Such drugs as neuroleptics, which are used to treat schizophrenia, as well as some antidepressants and Reglan (metoclopramide), which is frequently used in treating a wide variety of gastrointestinal ailments so common in the older population.

Mr. Kryhoski had taken no medicines except those he was on for his Parkinson's. He'd never been sick. He'd never missed a day of work, he reminded me.

I then asked him about industrial solvents. They sometimes caused neurological problems.

He'd had no significant exposure that he could recall.

I got more specific. I asked him about toluene, which can cause brain damage. One more cause off the list. That left carbon disulfide; he'd had no exposure to that either.

Had he ever had encephalitis?

"What's that?" he asked me.

Brain fever, I explained, a fever associated with loss of consciousness or confusion.

He hadn't ever had encephalitis. He'd never missed a day's work. Hadn't I been paying any attention to his answers?

"Idiopathic Parkinson's disease," I wrote in my chart. Strictly speaking, "idiopathic" means the disease is self-generated, caused by the brain itself without any external factors. Although physicians today no longer believe that Parkinson's disease would occur without virtually any cause, the term "idiopathic Parkinson's disease" is still widely used today throughout the neurological world—in clinical practice, in research reports, and in scientific publications. "Idiopathic" has now come to represent "plain old-fashioned" Parkinson's disease, the disease that Dr. James Parkinson first described in 1817, one not caused by encephalitis or carbon monoxide poisoning or exposure to antipsychotic drugs. Peter Kryhoski seemed to have plain old-fashioned PD, or so I thought.

I then took a history of each and every one of his symptoms, as well any side effects he'd experienced from the medications he'd been taking. I went on to examine him and confirm the diagnosis. There is no blood test for PD. No brain scan, neither a CT scan nor an MRI. A patient has PD because he has the signs and symptoms of PD and no signs or symptoms that are not consistent with PD. Diagnosis is an exercise in plain, old-fashioned clinical medicine. This clinical exercise is the one thing I like about diagnosing PD. And Peter Kryhoski exhibited the four cardinal signs of PD: rigidity in his arms and neck; a tremor of both hands; a slowness of movement, which I'd seen as soon as he struggled to get out of the chair in the waiting room; and a decreased ability to maintain his balance. Four out of four and nothing that did not fit. He had Parkinson's disease.

I saw Mr. Kryhoski once every two or three months to track his disease and regulate his medications. I always saw him on Tuesday mornings, and he never missed an appointment. He never even got there late. After two years of treatment, his disabilities and the signs and symptoms had dissipated and he was much better than he'd been when I first saw him. In a progressive disease such as Parkinson's, success of treatment is often relative. Holding your own is a level of success. Improvement is certainly better, and maintaining that improvement is better yet. I was very pleased with how Peter was responding to treatment.

So was Mr. Kryhoski. "Call me Peter," he insisted.

"Everybody does. Can I ask you a favor, Doc?"

"Of course, Peter," I said.

"Will you see my best friend?"

"What's his problem?"

"Same as mine, only he's worse. Of course, he doesn't have you as his doctor."

Usually a taciturn man, Peter spoke on and on about Eddie Mraz, his best friend. They had gone to high school together and then on to work at the same job. Eddie's father had also been a welder all his life and had got both his son and Peter into the union. Peter and Eddie had another friend, Sam Rausch; they'd all been inseparable friends since high school, like the Three Musketeers. One for all, all for one.

"And I'll be D'Artagnan," I quipped. "Or perhaps Dumas."

"Now, Sam's not sick. He's fit as a fiddle. Of course, a bunch of us down there are."

"Down where?" I asked.

"At the union, of course. The Welder's Union."

I was suddenly interested. "How many men there have Parkinson's disease?"

"About fifteen."

"How many members are there in the union?"

"Just under seven hundred living members." Peter should know; he was the secretary ever since he retired.

Fifteen out of seven hundred had PD—that was over 2 percent. Parkinson's disease hits about 1 percent of the population over the age of fifty. A variation that small, from 1 percent to 2 percent, could happen by sheer luck. Bad luck, true, but still only a random distribution with Peter and his friend getting the short end of the stick.

"How many of the members are over fifty?"

That was another fact that he knew as secretary. "Less than half."

"About how many?" I persisted.

"Just over three hundred."

That changed the odds. That made the overall prevalence of PD fifteen out of three hundred, not fifteen out of seven hundred. That came out to 5 percent. The stick was getting a lot shorter. Five percent, five times the average, was unlikely to be the result of chance alone. Not impossible, but pretty darn unlikely.

I needed more information. I asked Peter about the nature of his work.

"I always did a lot of hard welding."

"I never thought welding was easy," I said.

Peter laughed politely. "That's a good one, Doc."

I hadn't meant to make a joke and looked at him in confusion.

"No, no," he said in understanding. "Hard welding is a type of welding. You weld with hard rods, rods containing lots of manganese."

"Manganese? How much manganese in the rods?"

That took some thought on his part. "I think somewhere around 10 percent."

I too had to give that some thought. I wasn't even certain that the exact amount mattered. Only that there was manganese there and that he had been exposed to it. "And in welding there are lots of fumes," I suggested. Fumes meant small particles of manganese would be inhaled. How much manganese got into the body from the fumes? I silently wondered.

"And you breathed those fumes."

"All the time for forty-one years. I never missed a day of work. I still miss those fumes."

"And all the guys do that?"

"Pretty much so."

A world of manganese miners, I said to myself. Then added out loud, "That could have caused your Parkinson's."

"I don't believe that."

"Why not?" I asked. "Eddie's got it too."

"But not Sam, Doc. He's healthy as can be, and he breathed the same fumes. And besides, even the union lawyer says it's unlikely. I only got Parkinson's after I stopped working."

All so logical and factual and in a sense truthful, but in my mind, all of that logic and truth could well be wrong. And here was Peter and his union brothers, all of whom had worked with manganese and 5 percent or thereabouts had begun to suffer from Parkinson's-like symptoms. I became fixated on why only some of the welders would get the symptoms.

I asked Peter if he or his friends wore masks, which would decrease the inhalation of manganese particles.

"I always wore goggles or a hood," he told me.

"To protect your eyes?"

"Sure, the union made us do that. No one wants to get their eyes injured."

"But didn't you wear a mask over your nose and mouth?"

"Nah, never. Why would we have wanted to do that? Welding is a good, safe way to make a living."

I nodded. Maybe it was and maybe it wasn't. But one fact had emerged. He had breathed those fumes every day of his working life. Forty-one years of exposure to fumes made up of particles of manganese. And the particles could be microscopic—far smaller than those in dust. Smaller particles are more dangerous than larger ones. They are far more likely to get all the way into the lungs and be absorbed into the body than the larger particles, which get trapped on the way down and filtered out by the respiratory system. It is the small toxic particles in tobacco smoke that work their way down into our lungs. The small particles of asbestos. The small particles of manganese.

Miners with manganese-induced parkinsonism have abnormal levels of manganese in their bodies, much higher than those seen in the world of nonmanganese miners. This is what we would expect. But their manganese levels are no higher than those found in manganese miners who are neurologically normal and who do not have parkinsonism. How could that be? Why doesn't every one of the miners come down with the disease?

Manganese poisoning has long been known to cause a disease quite similar to parkinsonism. The disease was first reported among manganese miners in the back hills of Chile. The disorder had pretty much been restricted to miners who spent many years living and working in an atmosphere filled with manganese dust and particles. Year after year they breathed those elements and some of them got sick. Not all of them, not even a majority, just some of them. That was what allowed the mines to continue to operate. If manganese poisoning affected all the workers in a year or two, the mines would have had to change their operation in some way or close. But it was only a few workers each year that fell ill, a fairly low annual incidence of the disease. So most workers were healthy.

Why would only a part of the population acquire the disease, and not all? The prevalence of the manganese poisoning could go up to about 25 percent, but the incidence remained low. (In the study of diseases, *prevalence* refers to the number of patients in a specific population with a particular disease at a particular time, whereas the *incidence* is how many new cases of a disease occur each year.)

The illness that struck these miners had several distinctive features. It usually consisted of two entirely separate phases. The first phase was primarily made up of what we would call psychiatric symptoms, so-called manganese madness. This phase consisted of an entire range of psychotic behavior. Some miners developed acute mania, others developed depression or mood swings, and still others showed classic, hallucinatory psychosis complete with paranoid delusions. Such madness was usually easily recognized and diagnosed. After all, the manganese madness was part of the social fabric—more a folk term than a formal medical diagnostic label.

The mad miner was then kept out of the mines, and his madness cleared. This was not only therapeutic but contributed to the evidence that manganese was the cause of the madness. If the miner with phase I symptoms stayed out of the mines, he usually did well. The madness did not return, and he would not go on to develop Parkinson's-like symptoms, the second phase.

Not all of those returning miners got sick again. But often phase I of the disease was followed a year or two later by phase II, the Parkinson's-like symptoms. Many of the patients had the same tremor, the same stiffness or rigidity, the same slowness or bradykinesia, and a very similar if not identical loss of balance as in idiopathic PD. In addition, the miners often had other neurological problems not seen in plain old-fashioned PD, including muscle spasms that varied from brief jerks to massive contorting spasms. Those spasms plus the two separate phases of the disorder made it clear to every scientist who ever studied the issue that this was a distinct disorder. Idiopathic Parkinson's disease had no initial psychiatric phase and was devoid of such spasms. Hence, the miners had a specific disease, clinically distinct from PD and with a specific, known etiology or cause—manganese intoxication. Because it had many of the features of idiopathic PD but was also partly different, it was christened "manganese-induced parkinsonism."

Manganese-induced parkinsonism had several other interesting features. Once the parkinsonism begins, it was invariably progressive. Unlike phase I, ending all toxic exposure by leaving the mines once phase II had set in did not change the course of the disorder. No matter what the miner did, the disease continued to progress, much like its idiopathic counterpart. Since no long-term study of these miners has ever been conducted, we cannot know for certain if there were any other long-term effects of the poisoning.

Two conclusions did come out of these observations. First, manganese-induced parkinsonism occurs only in those exposed individuals who are susceptible to manganese poisoning. What factor or factors are the exact basis for that susceptibility to manganese-induced parkinsonism are not known. There are several possibilities. Lack of a mechanism of clearing manganese from the brain could be the culprit. That would cause the individual to develop higher brain levels of manganese following exposure than in his disease-free brothers. So could a relative inability to repair nerve cells once they are partially injured by manganese. The levels of manganese in the brain would be the same, but the damage is far more severe and progressive. Peter Kryhoski got parkinsonism. So did Eddie Mraz. But the third Musketeer, Sam Rausch, did not. Why? Differential susceptibility—whatever that means. Creating the label gives this phenomenon a form of scientific credibility.

The second observation is not so obvious and requires stepping back to look at the data anew. Assume we are all miners. Not all human miners get parkinsonism. In fact, most don't. Moreover, those who do develop parkinsonism have no greater manganese exposure, no more toxicity, no higher levels of manganese in their brains than those who don't. Thus it seems logical to conclude that manganese cannot be the culprit. This is our usual interpretation of the world. In epidemiology—study of the incidence, distribution, and control of diseases in a population—this is how diseases and their causes are approached. It's the method that usually leads to results. We investigate pockets of disease and look for factors that might explain those pockets. We look for factors that vary and disease incidence that correlates with those factors.

These days, those of us who spend our lives trying to understand Parkinson's disease accept the notion that it is not as idiopathic as we first believed. The evolution of our concepts originated with a group of young patients in their early twenties who developed severe and even fatal parkinsonism from a "tailor-made" drug they shot up. No one thought these patients had plain old-fashioned idiopathic PD. We had a pocket of young patients with a new disorder. An epidemiologist's dream come true.

The dream bore results. The PD in these patients was found to be due to a chemical they had injected into themselves called MPTP (methyl-phenyl-tetrahydropyridine). The discovery that a man-made toxin could produce a disease that clinically mimicked most if not all of the cardinal features of idiopathic Parkinson's disease was key to our understanding more about the disease. In the twelve years since that observation, more and more research has been directed toward identifying the one or more environmental toxins that cause Parkinson's disease. However, all the research has been unsuccessful and will probably remain that way even though it's all been based on the usual epidemiological approach.

The research has been conducted the way we conduct virtually all such research. It is a search for suspects, sought in a routine way, based on traditional approaches. It is more reminiscent of the concluding scenes of *Casablanca* than of innovative research: Ilsa and her husband are on the plane to Portugal. Major Strasser arrives at the airport intent on stopping the flight. Rick shoots Major Strasser. He is standing there with the smoking gun in his hand. The gendarmes arrive. They turn to Captain Renault for orders. "Round up the usual suspects," he says, so off they go, and why not? The usual suspects commit most of the crimes in Casablanca. The plane takes off, and Rick and Captain Renault walk off to join the Free French, a stroll of about a thousand miles.

So our research into what could cause Parkinson's has ignored many features of the disease because they eliminate almost all of the usual suspects and our usual way of evaluating them. These usual suspects, I remind you, weren't at the airport. They weren't holding a smoking gun.

Any hypothesis that one or more environmental toxins play significant roles in the occurrence of idiopathic PD must account for a number of observations. First of all, the toxin must have been present in the environment since before 1817, the year the disease was first definitely described. This eliminates all man-made organic pesticides and insecticides. They did not exist in the 1880s and 1890s when we know PD was a common disease. We can throw out all of the research that looks at such new pollutants.

Second, since PD occurs throughout the world, the toxin must be widely distributed. We can thus discard any research into toxins that are not widely occurring, such as manganese. We do not live in a world of manganese miners or even of manganese welders. In other words, the culprit must be ubiquitous, since idiopathic PD has been diagnosed in individuals residing in widely disparate environments.

Third, the toxin must not be increasing in concentration since there is no evidence that the prevalence of idiopathic PD is increasing. For example, today only 3 percent of Americans live on farms—a far cry from one hundred years ago. Yet this enormous lifestyle change hasn't changed the odds of developing PD; the rate of incidence is still one out of one hundred over the age of fifty.

Fourth, the distribution of this toxin must be relatively uniform since the comparative prevalence of idiopathic Parkinson's disease is fairly consistent and few if any pockets of the disease have ever been described.

Fifth, a key defining notion is that no matter what the concentration of the toxin in any environment, only a very small minority of those exposed ever develop PD. Virtually every other toxin has been discovered because of distinct pockets of disease. A group of people get sick. They live in one place. Or they all have the same occupation. Or they drink the same water. Or they eat the same food. Or they inject themselves with the same illicit chemical. Whatever. That is how the usual suspects are rounded up and the one culprit is found. This does not apply in the case of Parkinson's disease.

Finally, we must not forget that humans are uniquely susceptible to this toxin since we are the only species to develop idiopathic PD. The disorder has not been reported in domesticated species of dogs, even though dogs show Parkinson's-like signs when given MPTP.

Then how could PD be caused by a toxin? Easy. It all goes back to Darwin and to Johanson's Lucy: Species' struggle to survive and the survival of the most fit. For at least that long, human child-rearing practices have assured survival of a subgroup within our species that is uniquely susceptible to a toxin that is nontoxic to other surviving individuals. Good old differential susceptibility: some of us have it—some of us don't. Specifically, scientists believe they have located a part of the brain called the substantia nigra, where automatic movement is controlled. Automatic movements include rolling over, sitting up, obtaining an upright posture, maintaining balance in the face of postural threat, walking, and swinging the arms while walking. All of these are learned without being taught, without even being observed—since all blind infants learn them as well. These are among the motor behaviors primarily disturbed in

Parkinson's. The ability to carry out these "primitive" or basic motor skills seems to depend on the functioning of the substantia nigra.

Every ability of the nervous system, including these basic motor functions of the substantia nigra, is susceptible to genetic variation, and this variability is true for all species. Some children learn to walk sooner than others. Some learn to run faster, or to jump higher. This obvious variation in basic motor skills of humans seems to imply a genetic variability in the robustness of the substantia nigra. This variability could come about in any number of ways. One is simply variation in the actual number of substantia nigra cells; those brains with fewer cells would perform the tasks dependent on these cells less well. Or the factor could be variability in how cells reach out and target other cells (a process called arborization); those cells that arborize the least would reach the fewest target cells—and be the least successful at getting their substantia nigra messages through. Another explanation could be a variability in enzyme concentration or function, or in the ability to repair RNA or DNA. All of these properties of the substantia nigra are influenced by genetic structure.

However, all animals demonstrate this same genetic variability in basic motor skills, whereas Parkinson's disease remains a uniquely human condition. Why?

My first conclusion: *man is unique in assuring the survival of virtually all offspring over a wide range of robustness of the substantia nigra.* In any species in which survival of the fittest is allowed to operate in its usual sense, any deficiency in this system would be selected against. Slow learning or performing prey or predators would not be likely to survive in the world for very long, certainly not to the point of reproduction. Fitness in genetic terms is really reproductive fitness. Thus in all other species, the slow, late walkers would be weeded out in every successive generation. The slowest wildebeest is separated quite quickly from the herd and gets eaten. And not by the slowest lion on the plain. The faster wildebeests escape the lions and survive to reproduce. The slow lions get gored by the water buffalo. And so it goes. Yet this is not true for man and never has been.

The basic robustness of the substantia nigra in any individual is of course genetically determined. If only the faster wildebeests survive, the percentage of slower, less robust wildebeests will never increase in the next generation. There are no slow, fully developed wildebeests, or at least very few. The lions see to that. But slower children are not punished for their slowness. They are not eaten because of the lack of robustness of their substantia nigras. They survive to become adults and reproduce. And the resulting offspring also tend to be slow. Over hundreds and thousands of generations. This has two results.

The first is that the normal distribution curve becomes flat. The slow genes don't keep getting weeded out, so speed becomes irrelevant. So does robustness of the substantia nigra, which determines whether the child can walk well at ten months or fifteen months. Whether he or she can run at two years or three years or even six years. It just doesn't matter. The second result is that the slowest of the slow can over time give rise to individuals who are even slower. This process results in the birth and survival of humans whose systems are far less robust than any member of any other species. No wildebeest this slow, with such a low level of robustness, would ever be born. The fantasy of a very slow, or even cowardly, lion lives only in our imagination. Natural selection continues to prevent this occurrence in generation after generation in one species after another. Except in humans. For us, the variability in the robustness of the substantia nigra has been deselected, or randomized, for millions of years. It is not merely that the normal distribution curve for number of cells or for enzyme activity has been skewed (shifted) to the left; it has also been flattened and denormalized.

As a result of this denormalization, individual humans with nonrobust substantia nigra neuronal systems are born and survive and can live to a ripe old age. Then comes the down side: some ubiquitous toxin enters the bodies and brains of every species. But this toxin is not very potent. It is unable to produce disease in individuals with a normally robust substantia nigra. The wildebeest does not become parkinsonian, nor does the lion. Only humans. And only a small percentage of them, those with less robust systems.

Assume that all individuals are exposed to the ubiquitous toxin. This toxin is relatively ineffective and can kill only the weakest surviving cells and then only by a relatively slow process. Even humans born with only 30 to 40 percent of the usual number of substantia nigra neurons will not be afflicted. PD occurs when only 20 percent of a person's substantia nigra cells are still functioning. If, in a lifetime of exposure, the toxin can kill only 20 percent of the usual number of cells in only a small number of humans, then only those with the least robust systems—the fewest number of cells to start with—would develop PD.

This hypothesis then explains the uniquely human nature of PD and how a naturally occurring toxin with worldwide distribution could cause a disease that is restricted to only one species. It also reconciles the notion that PD is directly due to an environmental toxin with the observation that only scattered individuals within the same environment develop the disease. Furthermore, this concept explains the apparent contradiction in our understanding of the epidemiology of PD and its toxic etiology.

The primary requirement in developing PD is not the ubiquitous aspect or concentration of the toxin but the denormalization of robustness. The classic problem solving methods used for the epidemiology of toxic disease is useless in this case. No matter what the concentration of the toxin in any environment, only a small minority of those exposed ever develop idiopathic Parkinson's disease. The most interesting corollary of this is that the co-called toxin may not even fulfill the criteria for a toxin. In available amounts it will not be toxic to most humans. In this regard, it is worth recalling that manganese miners who develop manganese-induced parkinsonism have neither greater exposure to manganese than those miners who do not develop parkinsonism nor a greater body burden of manganese. Individual susceptibility must be the key.

I saw Eddie Mraz. He had Parkinson's disease, which had started before he retired and progressed far more rapidly than Peter's did. There was less I could do to help him. He became increasingly disabled and died of pneumonia about four years after I first saw him.

I continue to treat Peter Kryhoski. Sam Rausch, the one who was "spared," drives him to my office. But Sam never smiles. He has a masked face, one of the cardinal features of the slowness and lack of spontaneous movement associated with Parkinson's disease. He also has a little more difficulty getting up out of a chair than he did last year. And every once in awhile I see a very mild resting tremor in his left hand. Sam thinks he is normal, and who am I to tell him otherwise? One of the reasons the overall prevalence of PD is as low as it is is because no one examines the Sam Rausches of this world. Perhaps we're better off for it, because humans are going to keep getting PD as long as we don't kill the slower children, which I like to assume will be forever.

5

I NEVER READ A MOVIE I LIKED

.

The Architecture of Reading

LOVE MOVIES BUT I hate films—especially the ones you have to read. I'd reached this conclusion long before I was a neurologist or even considered what specifically I might do with my life. Yet many of my friends who read the same books I did, saw the same plays, and some even who like the same operas seemed to love such films. For this reason I was certain that this void in my aesthetic appreciation was an unfortunate idiosyncracy arising out of a childhood wasted watching baseball games instead of reading great books. Of an educational upbringing that left me believing that movies were what you did on Saturday afternoon when it was too cold to play baseball, too wet to play football, and your mom was utilizing the entertainment center to listen to the Texaco broadcast of the Metropolitan Opera.

Over the years my perverse response to foreign films with subtitles was reinforced every time I suffered through one. Especially to French films. Comedies, I'm told. And I hated them all—as an undergraduate; as a medical student; as a neurology resident learning about the brain, its functions, and its diseases; and finally as a full-fledged neurologist. I went to those movies with my friends and came out of the theater feeling I had somehow been gypped. I hadn't actually seen a movie. I had read some aborted text that flashed by my eyes just as I was trying to follow whatever action there was on the screen above the text.

None of that is true anymore. I am a changed man. Don't misunderstand me—I still dislike reading such films, but two things have changed. I no longer go; I have given up inflicting such torture on myself. And far more significantly, I now understand that my inability to enjoy reading my way through movies is not my fault. Nor my parents'. The simple neurological truth is that human brains are not made to experience movies in that way. They may well have evolved so as to allow us to listen to Wagner (hard as that may be for many of us to accept), or to laugh our way through a Mack Sennett two-reeler, or to hope that this time Ilsa and Rick will somehow fly off together on that last plane to Lisbon, leaving Captain Renault and her husband to combat the Nazis. But the physiological systems of our brains were never selected to read text at the same time we were trying to watch these events unfold. So the simultaneous intermingling of these specific functions just can't be done efficiently.

Once you start reading the text, you can no longer actually see the movie. This physiological reality may be unfortunate, but it is true. Dubbing works, which is why American movies are regularly dubbed throughout the world. But subtitles don't work. Even those on the side of the screen, like Japanese titles, don't work. And if you think they do, you're fooling yourself.

This reality is the direct result of the way our brains evolved. Even in literate societies, being able to read has never resulted in a reproductive advantage. And remember that fitness—as in "survival of the . . ."—refers only to reproductive fitness. In fact, the opposite is more often true. Reading is associated with socioeconomic success, which correlates with a decreased birth rate. Reading, then, is a biological disadvantage. Fortunately for us, literate societies and reading have not been around long enough to influence evolution as yet.

My personal issue with subtitles was scientifically reinforced when I encountered a particular patient.

I was asked to see Terence Hennessey because he could no longer read. The neurological term for this is alexia, meaning "without reading." I learned very early in the examination that Terence was a man who had always loved reading, ever since he was a young child. Apparently he was one of those children who read so effortlessly that he'd virtually taught himself to read before the age of four. From then on, he'd read voraciously. At sixty-four, he still read as much as ever, if not more. A professor of English at Loyola University, he read everything from Homer to Hemingway and beyond. He read in several languages, in addition to English, including French, German, Spanish (which he'd learned specifically so he could read *Don Quixote* in the original), and Latin. All in all, Terence Hennessey was the perfect product of a literate society. He read, therefore he was.

Until the day he stopped reading.

The night before, he'd gone to bed, and as was his habit, he'd read before he went to sleep. The book was Julian Barnes's *Flaubert's Parrot*. Not as significant a literary experience as reading Flaubert himself, he told me. But far, far better than *The Bridges of Madison County,* which he was also reading, in order to discuss it at a book club he chaired, one of the ways he supplemented his living.

Anyway, the last night he ever read, he finished *The Bridges of Madison County* and wrote down some notes. He picked up his well-worn copy of Marcel Proust's *Swann's Way,* but he was too tired to read any more. It was time to turn off the light and call it a night.

He slept as well as he ever did. Just like clockwork, he was aroused at two-thirty by the need to urinate. That had been happening for the last several years at almost precisely the same time every night. He had even talked about that with his physician in between discussions of Wilfred Owen and Siegfried Sassoon and the other poets who came out of the trenches of the "Great War." Terence was certain that he understood Sassoon far better than he did his nocturia. He seemed to take some pleasure in the fact that he had learned the correct term for urination during the night and even explained its derivation from the Latin. His kidneys functioned well. So did his bladder.

The problem was his prostate. It had been enlarging ever so gradually for a number of years. As it enlarged, it began to constrict his urethra, slowing the flow of urine and as a result preventing him from emptying his bladder completely. That meant that every three hours or so his half-emptied bladder would once again become full. Once his bladder became distended, it did what every well-behaved, distended bladder was supposed to do. It woke him up. He got out of his warm bed and walked into the bathroom and stood there trying to urinate. This was not as easy as it once was. The same constriction of flow that led to his bladder always being half full also obstructed the initiation of the flow of urine out of his bladder. He tried to overcome that partial obstruction to once again half-empty his bladder. And tried some more. He employed a Valsalva maneuver. He looked up from his hospital bed and smiled at me. "It's named after a great eighteenth-century Italian physiologist," he said.

"I know," I replied, not certain whether he preferred to educate me or was checking to be sure he had a physician who was already educated. It was, I decided, a toss-up.

He did not let my apparent knowledge interrupt the flow of his discussion. He was, I reminded myself, a professor of English who spent his life lecturing. This was his version of a conversation. I too was a teacher who lectured and taught at the bedside. Had I too fallen into such habits? No, I had my wife, who kicked me every time I lapsed into such digressions. So I let Terence Hennessey explain the Valsalva maneuver to me. It was designed to increase abdominal pressure and thereby help to empty the bladder. The maneuver is done by first closing the glottis (the space between the vocal cords) to prevent any air from leaving the lungs, and then trying to exhale against it. In the process, the diaphragm pushes down against the abdominal contents and compresses them, resulting in increased pressure within the abdominal cavity. This in turn pushes against the bladder and helps to force the urine past the obstructing prostate gland.

"So that is what I did," he concluded. "Sometimes when I do this I get a little dizzy," he added.

"Do you know why that is?" I asked him.

"No, not really. I teach literature, not human physiology," he replied rather sheepishly.

So I explained the process to him, knowing full well that before the day was over he would be explaining it to the medical students and nurses and residents who might wander into his hospital room and anyone else who would listen. The Valsalva maneuver also involves an increase in the activity of the vagus nerve. The vagus is the most important nerve in the autonomic nervous system, that part of our nervous system that acts primarily in an automatic fashion as a series of reflexes. It was his autonomic nervous system that had awakened him.

"I thought it was my bladder," he said.

"It was, but that message had to get from the bladder to your brain."

"I never thought about that. I guess that's what makes me a professor of English."

THE BLADDER IS controlled by the autonomic nervous system. Every parent is well aware of this. Even in a newborn infant, urine doesn't just dribble out at the rate that it is formed by the kidneys. It's stored within the bladder. Then as the bladder becomes distended, it sends a message to the spinal cord, and a message comes back to the bladder telling it to contract and to the sphincter, which is closing off the urethra, telling it to relax and that it's time to change the diaper again. A simple reflex. All mammals function in pretty much the same way. Dogs, cats, rabbits, humans.

And all mammals learn to suppress this spinal cord reflex.

Puppies do. Kittens do. Baby rabbits do. Toddlers do. That's where the brain comes in. The bladder fills. It sends its message to the spinal cord: "it's time to empty out." But once the brain has learned to inhibit the emptying reflex, the spinal cord no longer automatically controls it. The message gets relayed to the brain, and the brain then makes the choice: to empty or not to empty.

"So my brain figured it was time to get up and empty my bladder, ergo I woke up."

I nodded. I hoped the medical students would grasp his teaching as quickly as he had understood mine.

But Terence was not some twenty year old who, after three beers before bed, woke up because of his expanded bladder. A twenty-year-old man could simply stand there and allow his reflex to become unsuppressed and his bladder to empty. But Terence's bladder was hard to empty, and he experienced resistance to outflow due to his prostate.

"So I performed a Valsalva. But why did I get dizzy?"

"Because the autonomic nervous system works so damn well that it has never evolved. What it has gained in efficiency it makes up for by its lack of specificity."

I knew he was lost, so I stepped back and started again. The vagus nerve carries many messages to many places. As the abdominal pressure goes up during a Valsalva maneuver, the vagus nerve is also firing away. That slows down your heart rate, I told him. Your pulse rate becomes very slow. And it also lowers your blood pressure. "With that combination of slow pulse and low blood pressure, the blood flow to your brain is decreased and you may get a wave of dizziness."

"It was more than a wave," he said. "I often get slight waves, but this was a veritable kamikaze."

"So what happened?"

"I must have been Valsalva-ing for several minutes, and then very slowly my urine came out. Just like every other night. Not exactly 'Old Faithful,' but faithful enough. Then it came to a whimpering halt. I'm certain that it was only half-emptied, once again. A few more dribbles and I went back to bed." Back to his considerations of *Flaubert's Parrot*. He wanted to reread *Madame Bovary,* he told me, in French, of course. He hadn't read it in over ten years. In either French or English. How could that have happened? And he also wanted to read something else by Barnes, *A History of the World in 10¹/₂ Chapters*. Then he would read parts of *Flaubert's Parrot* so he would be all set to start out on the talk circuit.

Half the men in this country who were his age have the same problem with their prostate. Or worse. But as I listened to him, I realized that not one single hero in any major work of fiction I had ever read had to spend his nights wrestling with nocturia, urinary hesitancy, bladder retention, or any other attribute in the litany of symptoms comprising benign prostatic hypertrophy. So much for realism in literature.

But his bladder continued to wake him. At five-fifteen, his bladder was once again filled to overflowing. The urge again awoke him. Another trip to the bathroom, and once more it took a Valsalva maneuver to force "Old Faithful" to dribble its way into the toilet.

"Were you dizzy again?" I interrupted him.

He thought. "I guess so. I'm so used to a little dizziness that I hardly pay any attention."

"No kamikaze," I commented.

"No kamikaze," he agreed.

This time he hardly considered which of the two Dumases to read, father or son, before he was once again asleep. In no time at all, his alarm radio started to play. It was seven. It was turned to WFMT, the best classical music station in the country. They were playing a piece by Samuel Barber, his "Overture" to *The School for Scandal.*

I had listened to that same piece that morning while looking into the mirror in my own bathroom, seeing my entire face covered with shaving cream and then carefully manipulating my safety razor as I watched the pathways being formed within that lather until not a trace of white was left on my face.

"I prefer the Sheridan play," he said. (He was referring, of course, to the eighteenth-century play *School for Scandal,* which had inspired Barber.)

I didn't, but I merely said, "I can see why." Which I could. Language, the written words, were his life.

Terence then described how he went through his usual morning routine: another emptying of the bladder, entirely by reflex while the shower was warming up. A shower. A shave. Getting dressed. Making the coffee, and the toast. Retrieving the paper from the front lawn. And that's when he found out he couldn't read. He could see the paper; he knew it was indeed a newspaper. But when he looked at the front page, he was unable to compute the letters, or see them as forming words, which, in turn communicated sentences and meaning.

"Did you notice anything out of the ordinary?"

"No."

"When you looked into the mirror, did everything look normal?"

"Yes."

"And you had no trouble shaving? No difficulty manipulating the razor in your right hand?"

"No."

"And you were able to shave both sides of your face equally well?"

"Of course."

"How can you be certain?"

"I could see." He paused, following my train of thought and then added, "I also felt my face. I always do. To make sure it is smooth and the two sides were equally smooth. I felt with my left hand. So why couldn't I read? Why can't I read?" This last had more than a hint of desperation.

If only the medical students were as quick. He had told me what I needed to know. His right hand had not been affected. That is, his brain's left frontal hemisphere was functioning as evidenced by the fact that he could carry out movements requiring fine control such as shaving his face. This also implied that there was no visual loss in the central part of his visual field, the part we call macular vision at the center of the retina. Different parts of the retina respond better to different sorts of stimuli. The periphery responds better to movement and large shapes. The center, the macular part, responds better to details and colors, which allows us to tell one face from another, one letter from another, one word from another. To read.

That morning Terence could see his face clearly. He could also see his newspaper, the entire shape of it, and recognize it for what it was—a newspaper. He understood there were words on it that he should be able to read. If he couldn't read, it wasn't because his eyes could not see the letters or relay them to the visual cortex.

Was his problem really as specific as it sounded—restricted to an inability to read?

With most patients, it is only a few aspects of the examination that really matter. Sometimes only a glance is needed, or tapping on a specific reflex, and the rest is merely icing on the cake.

Where to start? His visual pathways were apparently intact. I checked just to make sure. I stood in front of him and made certain he could see and recognize objects and count fingers in each quadrant of his visual fields. No problem there, which confirmed that the visual images of the words on his newspaper had gotten to his visual cortex at the back of his brain.

Next I considered whether his speech centers were okay. If they were, then the front half of his language cortex was in fine shape. I already knew he did not have a problem with the production of speech. Patients who have a localized problem with speech output always put out less speech. They also make errors in word selection. And the category which subfluent patients have the most difficulty with is nouns, especially proper names. Terence Hennessey was anything but subfluent: He used a broad vocabulary and had no problem using the correct words. He easily found names ranging from Flaubert and Proust to Sassoon and Owen. For all these reasons, I knew Terence's speech output was normal. So it must be a problem with input.

Was the problem of input isolated? Could words be transformed into written output? I asked him to write a sentence.

"Any sentence?"

"'This is a beautiful day in Chicago,'" I said, handing him a pen and a piece of paper.

He wrote it without any hesitancy or misspellings. So I knew he could receive a verbal stimulus and repeat it as a written response.

Could he make up his own sentence?

He did, almost: "All things considered, I'd rather be in Philadelphia"—the epitaph on W. C. Field's tombstone. He had not created this sentence anew, but he had conjured it up intact from the language portions of his brain and written it out absolutely correctly.

He could spell correctly. Virtually every patient with aphasia has trouble spelling, especially spelling backward. Mr. Hennessey could spell "Flaubert" forward and backward, as well as "Stendahl" and even the names of those Russian novelists for whom no one can agree on the correct English transliterations. Spelling was the last formal test I gave him. There was nothing wrong with his linguistic brain. Chomsky would have been proud of him.

But he couldn't read.

My examination of him had confirmed that visual images were conveyed normally from his retina to his visual cortex. The words got there. And whatever got into the language systems of his brain was processed normally. Except for the written word. Ergo, the problem was that words could not get from his visual cortex into his speech centers. That meant his problem was localized within a pathway his brain had selected and reinforced since he was four years old, when he first started to read.

Reading is an artifact. An aberration. To use Stephen Jay Gould's term, it results from a spandrel. Gould borrowed this word from architecture after observing the Cathedral of San Marco in Venice. Spandrels are structures needed to mount a dome, such as the dome of San Marco's, on rounded arches.

Once in place, spandrels take on an artistic function of their own; they become an integrated part of the whole, as well as part of a decorative tradition. It is this feature Gould wants us to consider when applying the term to evolutionary biology. A spandrel would be a structural or biochemical necessity originally selected to serve a specific function, but which then is commandeered for some other entirely different function. It is not that reading is a spandrel; rather, reading is a by-product made possible by spandrelization. Our brains did not change in structure in order to allow reading; indeed, reading hasn't been around that long—less than six thousand years. Once upon a time, no one read or wrote. Then one fine day, one smart Sumerian started to write by drawing cuneiform symbols in clay, and suddenly the brain had to learn how to read. But in the brains of people who have been living in a nonliterate society nothing changes after they are introduced to reading. By the turn of the nineteenth century, most of the over five thousand languages in the world had no literature at all and were not written down. Yet people raised without a written language learned to read, and their children learned to read as easily as ours do. Reading, then, is a function most of our brains can do and do quite well, but unlike spoken language, it is not something they were designed to do.

We were, however, designed to see, and that design is not simple or straightforward. Light passes through the lens. And the fact that the eye has a lens is a pivotal factor in the evolution of our brain. Because of the lens, an object entering our field of vision from the right side of the outside world (the right side of our field of vision, or right visual field) hits the left half of our two retinas. In order for our brain to interpret these two images as only one object, they must get back together. The image on the left side of the right retina and that on the left side of the left retina must reach the same place in the same hemisphere at the same time. How does that happen?

Through a partial decussation, a partial crossing. The images from the left half of the retina of the right eye cross over to join those from the left half of the retina of the left eye and find a home in the visual cortex of the left side of the brain. In this way, the right half of the outer world is seen by the left half of our visual cortex. The images come together at the same spot from the right visual field to the left half of both retinas to left visual cortex. The same is true for the other side: the pathway runs from the left visual field to the right half of both retinas to the right visual cortex.

This design applies to both peripheral vision and macular vision. The partial decussation, which is the spandrel, probably evolved first for peripheral vision and the need to turn our heads and eyes to follow that object in our visual field which had been displaced to the wrong side by that lens. And it is the same for macular images, which are crucial to the recognition of letters and words—to the entire process of reading.

What has all of this got to do with Terence Hennessey and his inability to read a newspaper that morning?

Language is ordinarily a function of the left cerebral hemisphere. Reading is a function that depends on integrating the visual cortex with the speech areas of the left hemisphere. This is the pathway that gets spandrelized. Somehow the brain must get those symbolic images into the language cortex to be understood. But which visual cortex is involved—left or right? That doesn't seem to matter, a fact which, as will become evident shortly, shows that reading is based on one or more spandrels. Not a Rube Goldberg solution, but not as simple or elegant as speech either.

We read with only one visual field. Or at least, with only one at a time. How do we know that? Back to clinical neurology.

Recall that the left half of the visual field goes to the right visual cortex. If a patient loses all function of the right visual cortex, he can no longer see the left half of the external world. Such a patient would not see the entire left side of his face when looking straight into the mirror in the morning. If he were out driving, he would not see cars coming into the intersection from his left. Not a good situation. Far more dangerous than having vision in only one eye. That one eye, because of partial decussation, sees both halves of the external world.

Can the patient with vision in only one eye read? If he reads English he can. English is read from left to right. We read with our leading visual field, so each letter and word is read as it enters the right half of the visual field. A patient with the loss of the left field of vision, a left hemianopsia, has no loss of function within the right visual field, including the macular portion. So each letter and word is seen in the macula of each eye and relayed to the brain in a normal fashion. The patient can read.

But one problem remains. Peripheral vision is what we use to find objects as they move into our field of vision. And either the object or our visual field can be what is moving. The latter is how we find where the next line begins. This time it is the left visual field that gets used. Patients with a left hemianopsia have difficulty finding that starting point. They learn to use a finger pointing to the left end of the next printed line so their remaining visual field can locate the starting point.

Reading is managed by the design of lens to cortex to field; the partial crossing of the right visual field to the left hemisphere has been spandrelized for reading. A perfect design, except for one little fact. Reading doesn't have to involve the left visual cortex. For as George Bernard Shaw put it and as reworded by Lerner and Lowe, "And the Hebrews do it backwards, which is absolutely frightening." Backward at least from the perspective of a reader of English or any other Indo-European language. Hebrew is read from right to left, and because of this, Hebrew is read with the left visual field, which goes to the right occipital lobe and must then find its way across the brain, another decussation, to get access to the speech and language areas, in the left hemisphere. If someone reads Hebrew, a left hemianopsia does not prevent reading; it merely makes it harder to find the next line. It is a mirror image of what the same injury to the brain does in someone who has learned to read a language that goes from left to right. Clearly there was no one design that evolved for tying together reading of the written word with the language system.

It gets more complicated. Many early languages were not read just left to right or right to left. They alternated. One line was read from left to right and the next from right to left, a system named after a Greek word meaning "as the ox plows the field." This system required the visual message of the first line of text to get first from one visual cortex to the language part of the brain, and then the images of the next line to get from the opposite visual cortex to the same destination in the language centers of the brain. This was not a case of partial decussation but of alternating decussation. It may have eliminated the problem of finding the starting place, but the added complexity had to be a hindrance. As a result, all horizontal writing and reading evolved into a single pathway, either right to left or left to right.

It's more complicated still.

Language centers are not always in the left hemisphere. They can be in the right, especially in an individual who is left-handed. But left-handers still read with the same visual field. So a left-hander learning to read English must learn to decussate from the left visual cortex to the language areas in the right hemisphere, while one learning Hebrew must learn not to. And all of this occurs without any conscious effort on the part of the reader. (The optic nerves, of course, continue to do what they always do, crossing over half of all the images so that the right half of the visual field gets to the left visual cortex and vice versa. It is the brain that has to learn what to do next.)

The brain just does what comes naturally, except what comes naturally is different in different languages and writing systems. And what comes naturally is even different in the same brain for different languages. The same starting place in the left visual cortex of most individuals reading English has to find its way to the left hemisphere. But if they are reading Hebrew, it starts out on the wrong side, and if they are right-dominant, both of those starting places have different final destinations.

And if the language is Japanese, another subtle, but real anatomical difference shows up. Japanese is read from top to bottom with the lower half of the visual field. Our reading spandrels have another complicating feature. Each half of each visual cortex—the half on the left hemisphere representing the right half of the external world, and its counterpart on the right hemisphere representing the left half of the external world—is divided by a deep fissure, or sulcus. This fissure is so important that it even has its own name, the calcarine fissure. Images from the top half of the world end up below the fissure, and those from the lower half above the fissure. If you are reading English, you are usually reading below the horizontal meridian. The same for Hebrew. That is why the reading lenses in bifocals are set below the horizontal meridian. That was why good old Ben Franklin labeled them that way. The images all languages read from side to side are above the calcarine fissure. Only languages like Japanese that read up and down differ.

And what of the Jewish child learning both Hebrew and English at the same time? Right to left, versus left to right. Which system does he learn? Both. And all more or less automatically. The power of a spandrel should never be underestimated.

How does the brain do this, design the right pathway? It doesn't design or construct anything. It *selects*. In the young child learning to read, the images of the letters and words streaming out of the four quadrants of the visual cortex—or, more properly, out of any one quadrant of the visual cortex at a time—don't get immediately channeled into one small pathway. They get relayed around the brain, to the left hemisphere speech areas as well as to the same inactive areas in the right hemisphere. And only those pathways that get reinforced through use become the predominant ones. Like the spandrels of San Marco Cathedral—they start as a support system and end up part of the overall design.

The choice of the pathway is once again an example of selection, one that is most efficiently done early in life, certainly before the end of adolescence. Pathways can most easily be switched during that period, when the speech center can decussate for individuals with acquired aphasia who change dominance for speech. At this stage, these patients can also switch the final location for the image of the written word without any particular difficulty.

I told Terence as much as I could. He had had a stroke. Just a small stroke, involving a small area of the brain—not more than a cubic centimeter at the most—one small tract of neurons. But it was the tract connecting the visual cortex with which he read, to his speech areas.

He asked the usual question: why had he had a stroke?

As usual the answer was always unsatisfactory. It always is. Why he was prone to having vascular disease of the brain was easy. He had diabetes. Over the years, that causes atherosclerosis (a buildup of cholesterol deposits) in the arterial system of his brain. High blood pressure does the same. The two most important risk factors for stroke, and he had them both.

But why that particular night?

High blood pressure not only causes atherosclerosis to clog up arteries and decrease the flow of blood through those arteries; it also destroys the muscular coat of the blood vessels of the brain. The blood vessels become rigid pipes which lose their reactivity. Normally when your blood pressure goes down, the muscular coat of the blood vessels in the brain relax decreasing the resistance to blood flow, so the blood supply remains steady despite the loss of pressure. But not for Hennessey—not that night.

It went back to the Valsalva maneuver: Increased pressure in order to urinate. Decreased heart rate. Decreased blood pressure. Less blood trying to force its way into his cerebral circulation. His vessels could not relax. The flow to his brain decreased. He had a stroke.

"All because of my prostate," he said. He had understood.

"Plus your hypertension and diabetes."

"But why that particular one cubic centimeter of my brain?"

"Some bad luck." Not very comforting that. But the truth. We don't know enough to understand why certain arteries get clogged more in some people and other arteries in other people. That was his one artery that was most vulnerable.

"Why that artery?" he reiterated.

He could have lost his ability to use language all together, resulting in aphasia. Or lost the use of half his body.

Hemiplegia. Or right hemiplegia and aphasia together. This combination could be caused by damage to not much more than one cubic centimeter, as long as it's the right cubic centimeter. Or rather, the wrong one. Many times I have heard doctors tell stroke patients how lucky they were, that it could have been worse. In a sense they are correct. But they are also wrong. No one is ever lucky to have a stroke.

"What do we do now?"

"A CT scan to help localize the stoke."

"But you already know exactly where it is."

That was true. That was what clinical neurology is all about. But the scan would define the precise location and the exact size of the stroke, which might help in giving more accurate prognosis. If the lesion was next to the tract and the tract was merely swollen, he would do well. If it was centered in the tract he might not.

"And other than that?"

We would try to figure out what we should do to prevent further strokes.

"But for this one?"

"The elixir of time," I said. "All strokes shrink over time. More of the brain stops working initially. As the stroke recedes, the deficit also recedes. We'll have to wait and see."

Neither of us found that very satisfactory.

The next morning I looked at Terrence Hennessey's CT scan, which indicated classic acquired alexia (loss of the ability to read) without aphasia. The scan showed a small area of infarction (poor blood supply) deep within the white matter (nerve fiber tracts) of his left hemisphere placed just where it had to be to block the flow of information from his left occipital lobe to his speech areas. It correlated perfectly with his history and examination. Clinicians generations before could have identified precisely where his lesions were located without the technology. The bad news was that there was no real swelling to recede. His injury was just where it should have been from a physiological viewpoint (that is, to match his symptoms), but just where it shouldn't have been in terms of his prognosis. Had it been a few millimeters to one side or the other and surrounded by swelling that would go away over time, he would recover his ability to read. But that was not what I saw in the cards—or on the scan, I should say.

Each day I checked in on him. We talked about various writers. From Sherwood Anderson to Stefan Zweig. There was little else I had to offer other than to listen. And his reading never got any better. We talked about listening to books. About learning Braille.

He seemed surprised that I would think his brain could learn Braille. I told him that there was no reason why he shouldn't try. After all, reading was not what our brains had been designed to do. When the Sumerians started to write on clay in cuneiform, the images their reeds produced in that wet clay could be seen with the eye—true reading—and once the clay was baked, the symbols could also be felt with the fingers. The brain was equally capable of performing both activities. The Sumerians opted for the former, but not on neurophysiological grounds, that is, not because the brain found the latter more difficult. Reading just kept the hands free to do other things. Had the Sumerians opted for the other solution, a form of Braille might have been invented. Then it would have been bilateral amputees who had to adapt by learning to read visually, while the rest of the world absorbed great literature through their hands. A touch of hyperbole, I readily admit, but it convinced Hennessey. After all, once he learned Braille, he could read Flaubert in either English or French.

The next day was Terrence Hennessey's last day in the hospital. He wanted to talk about Nikos Kazantzakis and a movie based on *The Greek Passion*. That was Terrence's favorite of all the novels of Kazantzakis. I preferred *The Last Temptation of Christ*. The movie was called *He Who Must Die*. Terence told me I should see it.

"Do the movies have subtitles?" I asked him.

"Of course."

"Then I'll pass."

"Why?"

"Because," I explained, "I've never read a movie I liked."

That, he assured me, was merely an idiosyncracy of mine, the legacy of a youth misspent rooting for the White Sox instead of reading Dumas. I was an adult now. I had to overcome it.

"I can't do that," I told him.

"Why not?"

"Because it isn't an idiosyncracy. It's a neuroanatomical reality." And suddenly I understood. It wasn't exactly an epiphany. I had stared at this revelation ten days before in radiology and been so concerned over my patient that I had not even understood what I was seeing. Not that there was anything more that I could do for him. In the intervening ten days, my unconscious brain had put it all together, and the entire explanation came forth as if it had been something neurologists had known forever. They hadn't. I hadn't. It had taken Terrence Hennessey to teach me.

I talked it over out loud with Hennessey, who was a great sounding board. His disorder demonstrated how the brain reads. Visual images from reading have to be recognized in detail. They have to be perceived in the macular area of the eyes, projected to the visual cortex in the occipital lobe, and then passed to the speech cortex in the left hemisphere. From macula to occipital lobe to left hemisphere. That much he followed, as we had talked about it before. Recognizing the specific features of individual faces also starts out in the macula and again goes to the visual cortex. Telling one nose from another is much like telling one letter from another. But then the information has to go from the occipital lobe to the *right parietal* lobe, all the way over on the other side of the brain, not to the speech areas of the left hemisphere. Our brains are not designed to do both feats at the same time. Sequentially yes, but not simultaneously. So if you are busy reading a series of words on the bottom of the screen, you can't really see what the faces are doing; you can't register their expressions or other specifics.

"I know that the actors are acting when I read a movie," I concluded. "I just can't do both at the same time."

"I watched *He Who Must Die* three times before I appreciated how good it was," Hennessey admitted softly. "And how wonderful an actress Melina Mercouri is."

"And you already knew the story."

"Of course."

"I always see movies at least twice before I review them," he added.

"Most of us don't have that luxury."

"But you should see it, at least twice."

I SAW TERRENCE Hennessey a couple of times after he was discharged from the hospital, more to monitor his progress than to anything else. He learned to read Braille within a few months and read it quickly and efficiently, although he still missed the feel of good leather bindings and the smell of old books. There are no old Braille books bound in leather. But he was able to read *Madame Bovary* again in both French and English. Then two years later he called me in a panic. His ability to read Braille was deteriorating. It was his diabetes. He was developing a mild neuropathy and . . .

I should have foreseen this. One of the first problems in diabetic neuropathy is loss of two-point discrimination. You touch a patient with the sharp end of two pins and ask him to tell you how many points he feels. A normal person can distinguish two pins on his fingertip even if they are only a millimeter or two apart. Not diabetics. They can't tell two from one until the pins are four or five millimeters apart. And the dots in Braille are closer together than that. What could he do?

How about listening to books?

The last time I saw Terrence Hennessey was when I ran into him at the opera. Puccini's *Tosca*. "I'm reading again," he told me. Bialik.

"In Braille?" I asked, wondering if he had found books with Braille imprints large enough that his slightly diminished two-point discrimination would not be a problem.

He shook his head. "Bialik."

Chayim Bialik, a Russian Jew, was the first modern writer to write in Hebrew in some two thousand years. He began the

Hebrew literary renaissance two generations before Israel came into being. Of course I knew Bialik.

"You should read his poetry. It is quite beautiful."

"I'm surprised you could find it in Braille. It's hard enough to find in English, much less in English published in Braille."

"Who said I was reading him in Braille? I said that I was reading him, really reading him. Not in Braille. Not in English. In . . ."

"The original Hebrew," I blurted out interrupting him. Of course. Why hadn't I told him to do that? Hebrew reads from right to left. "Backward." With the left visual field that decussates partly to get to the right visual cortex and then to decussate to the left language areas. A decussation which had not been destroyed by Terrence Hennessey's stroke. "You learned Hebrew."

"How else should I read Bialik?"

"Touché."

"I have also been reading Agnon. Have you read him?

Shmuel Yosef Agnon was Israel's only winner of the Nobel prize in literature. "Not enough. And only in English."

"And Amos Oz and Yehuda Amichai."

He was way out of my league.

"I even found a Hebrew translation of Stendahl."

I could almost feel the rapture in his voice. Flaubert could not be far behind.

"And now I'm working on mastering Yiddish. That would open up another entire world of literature for me."

"You'd better be careful there," I warned him.

"Why?"

I told him a story Marjorie Guthrie, Woody Guthrie's widow, had told me. She was the granddaughter of a famous Yiddish poet who spent every evening of his life reading from his own large collection of Yiddish books. She had no idea what they were, but they had given him so much pleasure that, after his death, she wanted to share them with others. She gave the entire collection of over four hundred volumes to Brandeis University's library to build up its collection of Yiddish books.

"And they all turned out to be pornography," I finished.

"That's why I need to read more than just Hebrew," he said and walked away laughing.

I DID SEE *He Who Must Die*. Three times. The first time I read the subtitles. The second time I half-read them. The third time I finally saw the movie. It is a masterpiece. Almost as good as the book, despite the English translation and despite being in Braille. Or so I'm told.

6

—

ONE OF THESE THINGS IS NOT LIKE THE OTHERS

·

How Literacy Changes the Brain

I MET SAMUEL HAIRSTON several days after he had arrived at the hospital in a coma. Mr. Hairston was a jackhammer operator, and this was not the first time he had fallen into a coma, a constant occupational hazard during winters, when he operated his jackhammer indoors, in spaces that were poorly ventilated. The comas were due to the frequency of carbon monoxide poisoning. More than once he had been pulled out of a building in this state. Often he woke up quickly after being dragged unconscious into fresh air, and needed only to take the rest of the day off and get some rest. It was only if he didn't wake up that he was ever brought to a hospital.

One of the residents taking care of Sammy was worried that after he had been revived, he'd lost a great deal of cognitive function, that both his overall level of intellectual function and his memory had been altered. The resident asked Sammy how many times he had actually fallen into a coma.

Half a dozen times or so, Sammy said.

All gas-driven engines can produce carbon monoxide poisoning. When it gets into the bloodstream, it causes trouble. Carbon monoxide is not just carbon dioxide without one atom of oxygen; it is a far more active substance. Whereas carbon dioxide simply dissolves in the blood and gets carried to the lungs to be diffused out of the body, making it essentially a passive substance, carbon monoxide binds to the hemoglobin that is being pumped through the lungs. In this way, carbon monoxide actively displaces the oxygen that hemoglobin was designed to pick up in the lungs and deliver to the various parts of the body.

It's when the concentration of the hemoglobin-carrying carbon monoxide (also known as carboxyhemoglobin) gets too high that the real problems begin. When the hemoglobin is all tied up with carbon monoxide, oxygen itself cannot be bound and transplanted to body tissues. Consequently, those parts of the body that are most dependent upon the moment-to-moment delivery of oxygen begin to suffer from its decrease. This lack of delivery directly affects the brain, making it unable to function and after only a few seconds causing a person to lose consciousness. Once this occurs, the kidneys are still working fine. And the liver. And the heart. The lungs would also continue to function if the brain would only drive the diaphragm and the other respiratory muscles. But as soon as the brain stops working, the respiratory muscles are no longer directed to push air in and out of the lungs. This is how CPR (cardiopulmonary resuscitation) works; by mechanically forcing oxygen into the lungs and carbon monoxide out, the heart and the brain continue to function. If the individual is resuscitated immediately, no harm is done. But if the resuscitation takes too long, that patient can become severely brain damaged, or brain-dead, though many other organs can go on functioning for years.

That is how carbon monoxide poisoning works. The brain is the organ that is most likely to be permanently damaged. However, not all areas are equally likely to be affected. One of the most susceptible is a structure deep inside each hemisphere known as the globus pallidus. Injury to this part of the motor system can result in the same signs and symptoms found in patients with Parkinson's disease. People suffering from idiopathic PD exhibit tremors, rigidity, slowness of movement, and loss of control of postural balance, all due to progressive loss of cells in the substantia nigra. When these symptoms follow carbon monoxide poisoning, they are due to destruction of the globus pallidus. For this reason, patients such as Samuel Hairston are diagnosed not with PD itself but with carbon monoxide–induced parkinsonism.

The resident asked me to evaluate Sammy to make certain that he did not have *any* degree of parkinsonism. As I walked into the room, another neurologist was testing his intellectual function, going through the usual rigmarole: name the presidents backward beginning with Nixon.

"Who?" Sammy asked.

"How about mayors of Chicago?"

"Daley."

"Can you name any others?"

"Has there been any others?"

That was, I thought, a very good answer. Sammy had a distinctly Southern twang. I had no idea how long he'd lived in Chicago. Perhaps he'd only been here for ten years. Perhaps "dah Mayor" was the only mayor of Chicago he had known or even heard of. The neurologist went on to another subject—current events. Sammy had no idea where Hanoi was. Or who Ho Chi Minh was. Or Mao. The examining neurologist's questions continued, and Sammy's answers got no better. His performance on past events was equally bad. He didn't know what the Bay of Pigs had been. Nor the date of Pearl Harbor. He did remember that Kennedy had been shot and like many Americans even knew what he had been doing when he'd heard about it. But overall his performance had been dismal.

The neurologist then took out some pictures and told Sammy to look at them. He instructed Sammy to choose the photo that didn't belong to the group. The first picture showed a hammer. The second a piece of wood. The third a chisel. The fourth a wrench. Three tools and a hunk of wood. One of these things was not like the others. One of these things just didn't belong.

Hammer, wood, chisel, wrench.

Sammy Hairston studied the pictures for over a minute and then shook his head. "They are all the same. They all belong," he concluded.

The neurologist shook his head and turned to leave the room, obviously having concluded, sadly, that Sammy had suffered a severe degree of neurological damage. Before leaving, he told the students that the exam of intellectual function had been concluded, but they should stay and observe my part of the examination.

No sooner was the door closed than I set to work. "Where were you born?"

"Mississippi."

"When did you come to Chicago?"

"In '28. I was thirteen."

"Did you ever go to school?"

"Two grades."

"In Mississippi?"

"Yahsir." All one word.

Two years of schooling in a segregated rural school somewhere in Mississippi in the 1930s. Not exactly a great deal of formal education.

"Can you read?"

He shook his head. He could make his mark, but he rarely had to: he was paid in cash. His wife did all the reading and writing; she had gone up to fifth grade.

He never read a newspaper. So much for the Bay of Pigs, and the date of Pearl Harbor.

"The news is on the radio, and TV," one of the students reminded me.

"You watch much TV?" I asked.

"Just baseball. I love baseball. Always did. I played a few games in the Negro leagues for the Monarchs."

The Kansas City Monarchs had been one of the major franchises in the Negro leagues before Branch Rickey and the Brooklyn Dodgers brought major league baseball kicking and fighting from the Middle Ages into the twentieth century, by hiring Jackie Robinson to integrate the sport.

I had the toehold I needed. "You a baseball fan?"

"You know it, a Cub fan and proud of it."

Usually those were fighting words, since I am a dyed-in-the-wool White Sox fan. I ignored the challenge and began my evaluation of this man who had been deprived of virtually all education. A man who was truly illiterate. For him there was no literature, no books, no newspapers.

"Who was Hack Wilson?"

"Our center-fielder. He hit fifty-six home runs in 1929. That still be unbroken."

He was almost right on both accounts. Hack Wilson had set the National League record by hitting fifty-six home runs in 1930, and that record still stood in 1974, when Sammy was my patient.

We talked baseball for another five minutes. There was nothing about the topic that Sammy Hairston didn't remember, or that he didn't remember correctly. From Hack Wilson in 1930, to Gabby Hartnett hitting a home run in the dark to win the pennant for the Cubs in 1938, to the Cubs winning the pennant again in 1945. He even remembered the one White Sox game he went to in 1948.

I knew which game he meant. "Cleveland," I said.

"I see Satchel pitch."

Satchel Paige had been the greatest pitcher in the Negro leagues. Everyone thought he was too old to get a chance to pitch in the major leagues when baseball had become integrated in 1947. By that time, Satchel Paige was forty-one. Who would hire a forty-one-year-old black pitcher to play in the big leagues?

Bill Veeck, who owned the Cleveland Indians, had signed Larry Doby to become the first black to play in the American League. Veeck also signed Satchel Paige, and Paige pitched in relief. He won six and lost one as Cleveland won the American League pennant. Paige and the Indians then went on to beat the Boston Braves in the World Series.

The night Sammy Hairston saw Paige pitch, the White Sox set an all-time attendance record. Satchel Paige shut out the White Sox. It had been the most important night of Sammy Hairston's life.

I looked at the now-confused medical students and said, "Memory, intellect, and judgment intact."

I don't think the students were convinced that this conversation about baseball constituted a neurological exam. Though I was satisfied that Sammy Hairston had none of the signs of carbon monoxide–induced parkinsonism. If you tested him on subjects he knew and cared about, he passed with flying colors. Remote events and persons, recent events and personalities—all were recalled and understood. Most of us don't accept an IQ test based on baseball as being valid. But it's as valid as any other. If your experience is baseball, then it's baseball that must be tested.

The students and I gathered in a small conference room to discuss our findings. But what about those tools? one of the students insisted. Sammy should have known the difference between a plank of wood and tools. It had been very clear to all of us that Sammy was unable to distinguish the one object that did not belong in the category. How could that be normal, the students wanted to know, when every four year old who watched *Sesame Street* could do it?

Alexander Luria and Walter Ong were the first to explain the apparent discrepancy.

In 1982, Ong, professor of Humanities in Psychiatry at St. Louis University, wrote a book, *Orality and Literacy,* in which he showed the difference between how oral cultures, those with absolutely no knowledge or use of writing, and how literate cultures, in which writing is deeply ingrained, handle information and knowledge. They don't do it the same way. That's what is so important. Literacy is so ingrained in our culture—at least, in the educated parts of our culture, in the "high" culture—that those of us who are literate don't even think about how brains might function in the absence of literacy. We are so darn arrogant that we call such societies "preliterate," as if all of evolution was a predetermined drive toward literacy. It wasn't.

What Ong was really saying, at least from the viewpoint of a neurologist, is that literacy changes not just how information is transmitted from one person to another; it also changes how the individual acquires and organizes it. And of course, this is not a societal difference in information processing. A society is nothing more than a conglomerate made up of individual brains. It's the way individual brains acquire and process information that becomes altered in a literate society.

"Sammy Hairston," I continued, "is not literate. For him, information and knowledge are oral concepts, oral experience. Knowledge is primarily experiential. He experienced baseball. He did not experience the Bay of Pigs. For us the world is primarily a literary experience, the written word. That makes his world and our worlds far different. More different than you might think. Ong stresses writing as the technology that has changed the entire process of discourse in the world. I would argue that for the individual brain it is reading not writing, that alters how the brain functions, how it classifies and thinks about the world and itself. One can read and yet never write. Reading changes everything. We depend primarily on visually related language. Sam Hairston depends on oral language. He only knows what he has heard."

I paused in my explanation to the residents. This was not something anyone else had discussed with these medical students. This topic was for linguists or people like Ong who were interested primarily in "the oral" versus "the written." (Ong in particular was interested in the evolution of religion.)

The residents did not seem convinced.

I tried to explain the differences between a culture that is primarily oral and one that is truly literate. Where to start? Much nearer to the beginning. Ancient Greece. The students had all gone to college. They had all read Homer. They had been introduced to Aristotle. There was the difference.

Homer was the product of an oral culture. Like all products of oral culture, his works required oral cues, mnemonic (memory) devices and other linguistic formulations to organize the story for his listeners. "The wily Odysseus," repeated over and over again, is a rhythmic formula in Greek. Homer's works also depended on the tricks of poetry itself: alliteration, assonance, rhythm. Poetry that could be recalled and that retold familiar stories. All of Homer was remembered history.

Aristotle was anything but that. His works dealt with classification and analysis. Not lists as lists, but lists as a way of organizing the world. Organizing and analyzing and making it possible to look ahead to further organization and analysis, to science. In an oral world, words are symbols and nothing more. They are evanescent, not finite. You can't build on them. But in a written world, words also become objects, their meanings are fixed. And you can do so many more things with objects. You can organize them, classify them, manipulate them in any of a thousand different ways.

"Sammy Hairston doesn't come from a literate culture, rather, from an oral one," I reiterated. "His brain has never received any linguistic input that was not essentially oral in nature, based on his immediate experience. He hasn't been exposed to systems of classifications or analysis. For him, language, memory—his entire intellectual apparatus—are set up to look back on what happened in his world and to organize it the way he experienced it. What the heck does Pearl Harbor mean to him? Nothing. He was probably never even registered with any draft board."

The medical students were beginning to understand.

Oral cultures, oral languages, I went on, have no dictionaries. Words have meanings, but they may not mean the same thing to everyone. Words acquire their meaning only from the way they are learned, from their connotative usage, from the sound, the inflections, the gestures—the entire human context in which the words were used. Since these words are not explained or defined in any formal manner, there is no way to verify a definition if one is given, except to ask someone else what that word means. There is no authority. Imagine a world without dictionaries. Each brain constructs its own vocabulary with its own nonverifiable set of definitions, except they are not real definitions as much as sets of appropriate word usages. There is no such thing as grammar: no classification into nouns, pronouns, verbs, adjectives, adverbs, prepositions, and so forth.

Enter Luria, the great early twentieth-century Russian neuropsychologist. He was the first to study word usage in oral cultures and to demonstrate that they organize information differently. He showed pictures of objects to illiterate individuals from oral cultures, and just like Sammy Hairston, they could not tell that a piece of wood did not belong with a hammer and a chisel, that it was not a tool. They had no conception of using classifications. Instead, like Sammy Hairston, the group thought situationally: when you build, you need wood, a hammer, and a chisel. The wrench is what doesn't belong.

"Because we in this room are literate," I continued, "we assume that the way we think is the only way the brain was designed to process information. We are dead wrong. Nonliterate brains in oral cultures, which have always been in the majority until the last few hundred years, organize knowledge in a far different way. Or rather, it is we, in the literate culture, who organize knowledge differently, and this form of knowledge has changed our behavior."

However, this transition from oral to post-oral culture did not happen overnight, nor in a single generation. Rather, the changes in thinking and behavior advanced in steps. At first, all information transmission was oral. Simple enough. Homer or Beowolf were the epitome, their epics learned from generation to generation by recitation. Then came the written word, which at first was used only by scribes to be read by other scribes. Then a few poets wrote for the theater. The Greeks and later Shakespeare wrote plays, which the nonreading populace then saw and heard performed. Then came the big change, the start of the revolution, so to speak: writing to be read by the masses. The society at large began to read, but people still didn't *have* to write. It is only very recently that the general population has been also taught to write.

But from a neurophysiological point of view, reading is the key advance.

Oral cultures have a simpler vocabulary. There are no dictionaries. There is no way to check the accuracy of any statement that is made. There can be no history. And no science. None at all. Little if any philosophy. Socrates could never represent anything more than discourse to be heard once and then recalled as best as it could be in as many different ways as there were listeners. Next came Plato and the written text. Plato wrote down Socrates' philosophy so that it could be studied and debated and learned and built on. Without Plato, Socrates would have been nothing more than a memory, a rumor, a myth.

If we look at the span of human evolution, from magical thinking to science, from myth to history, from the "savage" mind to the "domesticated" mind, we will notice how it's been affected by how we communicate and learn about the world. The oral world is made up primarily of simple declarative sentences with a simple vocabulary and depends on repetition—a world of sound bites. Yet it has been literacy that has been the basis of the evolution of human society for the last three thousand years or more. Make no bones about it. The industrial revolution is a by-product of literacy. As was the electronic age and, of course, the information age. The latter is an absolute triumph of literacy. Such evolutionary changes had nothing to do with DNA or Darwinism. They are effects of man changing his environment.

But the significance of reading doesn't end there.

Sammy Hairston represents one cog in the wheel of human evolution. Specifically, he embodies a particular step in humankind's ascent into a literary culture. It all goes back to the juvenilization of the human brain: the human brain matures by pruning cells, and the environment plays a pivotal role in that pruning. Sammy Hairston was never exposed to reading and the organizational components of literate society. His brain was pruned differently than yours and mine. Specifically, he learned to think, act, interpret, and respond to the external world in a way based solely on his experience. Not being able to read or write, his world was cut off from any kind of written symbolic communication, which in turn greatly limited his brain's conceptual input. The biology of his brain and ours is qualitatively different. Not as different as Genie's, or that of the Wild Boy of Aveyron, but different nonetheless.

Darwinism teaches us that the sons of great hitters often become great hitters because they inherited their fathers' predisposition to hit a baseball. But Jean-Baptiste Lamarck, a seventeenth-century French naturalist, saw things a bit differently. He believed that traits, such as the ability to hit a baseball, acquired during one generation could then be inherited, passed down through multiple generations, until the cultural evolution itself affected and finally altered the function of the brain. In other words, once there are great hitters, there is baseball, then professional baseball, and players' attributes as their brains get pruned. Before you know it there's the union, the baseball strike, and Mark McGuire breaking records. And so on.

This seems to be true in other cultures: Native Americans learned to be attuned to their "natural" environment. They could see and interpret tracks on the ground that the literate could neither see nor interpret. Were their brains any different? Not before the pruning of maturation had set in.

Will the information age change how our brains work? Undoubtedly. But how?

That is the sixty-four-dollar question. In this technological world of abstract, nonverbal symbols, our brains may be developing an entirely different, nonlinear way of organizing information. Perhaps we will use not lists and definitions to organize our world, but trees, trees covered with pictures and symbols to which none of them have words attached. Could we become more like native Americans following a trail than Abe Lincoln at the fireside reading a book? For now, it's all up in the air, as they say. And though I cannot foresee what the change will be, I'm quite sure that the brains of my granddaughters will prune different cells than my brain did.

7

THE MUSIC GOES ROUND AND ROUND

·

But It Comes in Where?

T HIS TALE STARTS in Milan. After consulting with a patient in Geneva, I was invited by a friend and colleague to accompany him to Milan, his hometown. As one of Italy's better-known neurologists and professor of neurology from Milan, Pietro Baldacci was trying to convince me that Milan was just a short trip away from Geneva. He suggested I could give a lecture in his department, and that evening he would bring me to a concert I would find interesting. It was a concert by a string orchestra conducted by Maestro Cesaro Rota.

Cesaro Rota was not a conductor whose name I recognized, but Professor Baldacci explained he was mainly known as an opera conductor at La Scala and at La Fenice, in Venice. La Scala was one of the world's great opera houses, but local Italian opera conductors were not known for their brilliance in front of an orchestra, Arturo Toscanini being the great exception. I was from Chicago. I'd grown up with the likes of Rafael Kubelik and Fritz Reiner conducting the Chicago Symphony Orchestra. I'd matured into a regular diet of Sir Georg Solti, Claudio Abbado, and James Levine. Some conductor of whom I'd never heard leading a string orchestra possessing an equal lack of renown held very little promise.

"My friend, it will be a concert you will never forget. I assure you of that," Pietro said. "It will mean more to you than all the concerts you have attended. Rota trained with Scherchen."

Now I'd heard Igor Stravinsky conduct Stravinsky, Aaron Copland conduct Copland, Leonard Bernstein conduct Mahler. Heard Reiner do Richard Strauss and Bela Bartok on the same program. I was more than skeptical. But I knew Hermann Scherchen was an outstanding orchestral conductor. He had been the first conductor to record a complete cycle of all ten (or rather nine-and-a-half) symphonies of Gustav Mahler.

"You will be astounded."

"Astounded?"

"As a neurologist, you will find it the most significant performance of your lifetime. You will come away with a better understanding of music than you have from any other concert you have ever attended."

A better understanding? Could this unknown really reveal things in music that had evaded the likes of Reiner and Otto Klemperer and even Toscanini? You must know that I have been an avid classical music listener since I was a small child. As well, my wife was trained in classical music and was a professional harpist; together she and I had regularly sought out live performances for over twenty years. This concert I had to hear. How could I say no to such a challenge?

So I stayed in Milan for an extra day. The lecture that I gave at the medical school went well. The chief resident presented a patient with Parkinson's disease, and I did my usual Visiting Professor gig. I used their patient as a takeoff point to talk about my own research into the mechanism behind the abnormal movements caused by so many of the drugs used to treat PD. It's a topic I had been studying ever since my first patient developed such movements, and it's a talk I can give without any need for rehearsal.

After the lecture, Pietro took me to a wonderful Italian dinner, and then we were ready to go to the concert. By then, my skepticism had returned. By delaying my return to Chicago, my wife had informed me, I would miss a concert of the Chicago Symphony. With Solti. Solti had studied with Bartok, she reminded me. Solti conducting Bartok.

"I'll be seeing Rota," I had told her.

"Who?" she asked.

"He trained with Scherchen."

Another name that had fallen into obscurity by the time she received her training in classical music. "Scherchen," I repeated.

"Bartok and the Mahler First," she countered.

I never won in these discussions.

The program that was handed to me as we sat down in the makeshift concert hall did little to reassure me. As far as I could make out, I was about to hear an evening of music for a pick-up string orchestra conducted by Cesaro Rota of La Scala fame. The program was made up of two Vivaldi pieces followed by Benjamin Britten's *Simple Symphony* and Ernest Bloch's *Concerto Grosso Number One for String Orchestra with Piano Obbligato*. All this instead of Solti conducting Bartok and Mahler. Not a good trade.

I'm not a judge of Vivaldi, so I will not comment on those performances. The response of the audience seemed reasonably enthusiastic. The Britten was pleasant enough, but not too enlightening, surely nothing that left me astounded. Not even a little bit amazed. As a conductor, Maestro Rota seemed stiff and uncomfortable, and he barely moved his right hand. For this I'd spent an extra night away from home?

Then came the Bloch and with it my first level of astonishment. The performance was nothing short of brilliant, in its own way as good as the old Kubelik recording with the Chicago Symphony Orchestra, the first long-playing record I had ever worn down to the point where I had to replace it, only to discover that it was no longer available. Maestro Rota gave an unforgettable performance.

When the concert was over, I asked Professor Baldacci if he knew the maestro personally.

Yes, they were friends.

"I would like to talk to him," I said. The Bloch had been great.

"You can't," he informed me.

"Why not?"

"You can talk to him, but it won't mean very much to him. Nor to you. He's aphasic."

I was not merely astounded: I was absolutely flabbergasted.

"He had a stroke," Professor Baldacci explained, "on his left hemisphere."

Rota had suffered a dominant-hemisphere stroke. That explained one aspect of his style of conducting, the lack of use of his right arm and hand. So those were not an element of his style but a result of the stroke. Maestro Rota had a mild residual hemiparesis (a slight lingering weakness on his right side) caused by his stroke. A permanent disability. After all, the left hemisphere does more than just act as the dominant hemisphere for speech. It also directs purposeful movements of the right arm and leg. It was common for a patient with a stroke resulting in aphasia to also have some right-sided motor problems—depending, of course, on the type of aphasia.

The conductor presented an interesting neurological case. Once again I reviewed my general rules for aphasia. If the problem is more toward the back of the dominant hemisphere, then the patient will have more trouble comprehending language than producing speech. This is the classic syndrome named after the German neurologist Carl Wernicke, who first delineated Wernicke's aphasia. Speech is still fluent but makes only limited sense, for the patient neither understands what has been said to him, nor filters or regulates what he is saying. The result is often called fluent aphasia, receptive aphasia, or just plain Wernicke's aphasia. Such patients usually do not have a hemiplegia (paralysis to one side of the body). Wernicke's area is not that close to the motor areas of the left hemisphere, and therefore it is easy to destroy one without injuring the other.

If the destructive lesion, the stroke for instance, is more toward the front of the brain than the back, the patient will have more difficulty producing speech than comprehending it. This is the classic Broca's aphasia. Such patients are subfluent or nonfluent. And since Broca's area is adjacent to the motor cortex, such patients often have motor problems. The usual result is a right-sided hemiplegia, which on recovery often leaves the patient with difficulty using the right arm and hand.

"Maestro Rota previously used a baton?" I asked.

"Most certainly," Pietro said, "Scherchen liked long batons, moving quite crisply."

So Cesaro Rota had suffered an aphasia with motor loss. The diagnosis was obvious, and all things considered, not that surprising. He must have had a classic, expressive, Broca's aphasia. He was subfluent but could still express or perform music. In many patients the ability to express speech and the ability to express or perform music are located in opposite hemispheres.

The conductor could still express music though he was apparently not fluent. "A Broca's," I proclaimed in my most professional tone.

Pietro shook his head. "That would be interesting, I agree, but not astounding."

"Then what?" I asked. By now we were across the street from the small concert hall in a rather crowded and noisy restaurant having coffee and dessert.

"A global aphasia."

I was astounded. Global aphasia meant that Maestro Rota had lost all of his speech functions, both his receptive and his expressive abilities. Wernicke's plus Broca's all rolled into one. For Maestro Rota to have lost all aspects of his language function and to have so much of his musical ability preserved was certainly nothing short of amazing.

Many patients who have strokes that cause aphasia begin with a clinical picture that looks like global aphasia, but as they begin to recover and the stroke shrinks in size, their aphasias gradually improve as each part of their brain reorganizes to maximize its functional capacity. Such patients do not have static deficits: their loss lessens as recovery progresses. Their global aphasia shifts to a nonfluent Broca's aphasia as language comprehension is regained. Often such receptive elements return almost to normal, which is what must have happened to Maestro Rota: from a global aphasia to a Broca's aphasia in nine months or a year.

Unlike the recovery from hemiplegia, which stabilizes within a few weeks, the recovery from aphasia takes far longer and usually leaves the patient with significant disabilities which impact on every single aspect of their lives. That must have been what had happened to Maestro Rota. From global language deficit to a more localized loss. From global aphasia with amusica (the Latin for "without music") to a Broca's aphasia without amusica.

The coffee finally arrived and with it a selection of biscottis. The brandy would follow. As I dipped my first biscotti into the coffee, I remarked, "He of course resolved into a Broca's."

Pietro gave me a disapproving glare. Was the dunking of biscottis a social *faux paux?*

Pietro carefully selected a biscotti and dunked it into his coffee. At least I had not embarrassed myself socially. Did that mean that I had embarrassed myself neurologically?

"When we first tested him," Pietro explained, "it was about ten days after his stroke. He'd been in the States conducting an opera. Puccini, *Madame Butterfly,* in Philadelphia. His stroke had occurred after the performance. He was hospitalized there for a week and then flown back here. At that time, my friend, he had a global aphasia. He was nonfluent. He uttered a few short phrases. And he had a mild right hemiparesis."

The professor interrupted himself to pick up another biscotti, dunked it into his cup of coffee, and simultaneously signaled me to follow suit as the rushing waiter brought more biscottis and fresh coffees.

"Cesaro," he continued, "could recognize single words. If we showed him a picture and gave him a choice of nouns, he could pick out the correct one. Both orally, if we said the word to him, and visually, if he read the words from a list that we showed him."

That meant that at least some comprehension had been preserved.

"But he could not understand sentences. Even simple ones. He could not even follow any but the simplest of commands. A one-step command he could perform correctly most of the time."

I knew the routine. You tell the patient, "Lift your hand," and he lifts his hand. "Open your eyes," and he opens his eyes. "Pick up the glass," and he picks up the glass. Simple comprehension and performance. Input and output at their most basic.

"Two-step commands were another thing. He did rather poorly."

Close your eyes, and lift your hand. A two-step command. Not very complex, but Maestro Rota had not been able to perform both tasks at the same time, or sequentially with a single set of instructions. Three-part commands had been a total disaster.

Not exactly as demanding as conducting Beethoven's Ninth. Or even one of the other symphonies without soloists and chorus. Without any words at all. There were no operas without words. Was that why he had reverted to conducting orchestras instead of opera?

A new plate of biscottis appeared; it was a large bowl this time, not a small plate, and it was overflowing with biscottis. Pietro smiled at the waiter, I smiled at the biscottis.

Pietro immediately returned to Cesaro Rota's initial deficits. Real comprehension of language was gone. "Finished."

"What did you do?"

"We admitted him to our rehabilitation unit. It is the best such unit in all of Italy. I am very proud of the work we do there."

I cringed. I knew what he was going to say: they had worked out a program of intensive speech and musical therapy which had worked miracles. Unfortunately the major miracles that most such efforts produce is directly related to the passage of time, which allows the brain to adjust to the new realities of its physiological reorganization. It's the elixir of time that is the secret to success, not formal speech therapy. "Pietro," I said, "I am sure your pride is justified."

He smiled back at me. "On the second day, even before our speech therapists had devised their complex program for successful rehabilitation, an old friend came by to visit Cesaro. Giovanni Lamberto had also been a musician. He too had gone to study with Scherchen. Scherchen, who had little use for bad musicianship, had thrown him out. So Giovanni had gone into medicine and become a radiologist. He'd opened his own radiology facility and had long since retired to count his money. Second-rate radiologists often get rich. And he was second-rate. I knew him quite well. He had the largest radiology practice in Milan and a wonderful box at La Scala. I would see him at the operas often. He never missed anything that Rota conducted. Strange, isn't it, that most first-rate musicians barely make a living?"

Italy was not all that different than the States.

Pietro continued, "He was shocked to see the condition of his old friend. Rota could recognize him but not much else."

"So Giovanni decided to play some music for Cesaro. He wheeled Cesaro into the sitting room near the piano and began to play for him. He played the Intermezzo from Mascagni's *Cavalleria Rusticana*. You do know it?"

Of course. It was Pietro Mascagni's most famous piece, his greatest hit.

"It was Giovanni's favorite piece of music. That was because it was the piece he'd played for Scherchen when Scherchen had ended Giovanni's musical career, which made Giovanni change his life and made him rich, successful, and powerful. He'd never even gotten through the entire piece for Scherchen before he'd been cast out on his rear end. Lamberto must have told me that story a dozen times. At least once every time Rota conducts *Cavalleria*. And you have no idea how often La Scala does

Cavalleria. Giovanni never misses a performance. For him it was like going to mass.

"So Giovanni sat down at our old piano and played the Intermezzo from *Cavalleria.* I suspect that Giovanni played it for Cesaro much as he had played it for Scherchen over forty years earlier. Giovanni was still the same musician he had been back then, but Cesaro was not. Cesaro had studied with Scherchen. When auditioned with Maestro Scherchen, he had not played Mascagni. He had played Mahler, Zemlinski's piano version of the Sixth Symphony. And Cesaro, too, had been interrupted by Maestro Scherchen—not by an order to give up music and do something else to make a living, but by an offer to become an assistant conductor with the maestro."

By now the coffee was cold. Neither of us had even touched a single biscotti from the fresh bowl.

"Giovanni played, and when he got to the exact point, the very note, where Scherchen had halted him and changed his life, he once again heard the words 'Stop! Stop! Murder! The music, the music.' It was Giovanni's personal epiphany being revisited upon him. But it was not the ghost of Scherchen. It was Scherchen's disciple, his old friend Cesaro, demanding that Giovanni stop killing the music. In short, subfluent sentences, Cesaro was very expressive. He might have had a global aphasia, but when it came to music, he knew what he wanted to say. And he could say it in a few words."

Giovanni stopped playing. But this time he did not end the performance. "He started again from the beginning. He played it more slowly this time. 'No,' Maestro Rota shouted again. And once again Giovanni pulled his hands off the keys. Giovanni could see his friend's disappointment and anger. There had to be something he could do to make his friend feel better. More alive. Less frustrated. To give him back at least a small part of his life. What? How could he put music, the creation of music, back into Cesaro's life? Giovanni moved the wheelchair so he could see Cesaro and play at the same time. Then with a single word he changed Cesaro's life.

"'Conduct,' Giovanni declared to his friend Cesaro.

"Cesaro sat there in the wheelchair and tried to lift his weakened right arm. It moved weakly and slowly. Awkwardly. Hardly at all. His right hemiplegia was less forgiving than his old friend. And far less adaptive.

"'Conduct,' Giovanni repeated.

"Cesaro tried again. His right arm was useless. Giovanni felt useless. He would play something else. Cesaro had always liked Richard Strauss and conducted a number of Strauss operas at La Scala. He also liked the tone poems. Perhaps Giovanni thought to himself, he could play a snatch from one of the tone poems. Giovanni could not remember any of them. Had he ever really known anything by Strauss that well? He'd never liked Strauss. Not even the piano piece Strauss had composed for the left hand or for Wittgenstein." Paul Wittgenstein was a great pianist who toured throughout Europe before World War I.

"To make the world safe for democracy," I said.

"Yes. And we were one of the democracies made safe by that experience. So we got Mussolini out of it—no matter. That is not the point, my friend. Wittgenstein had lost his right arm in the Great War. A tragedy for anyone, but for a professional pianist, it was a complete disaster. There was no music for one-handed pianists. But Wittgenstein had charisma, and his family had money."

"As my mother taught me," I said paraphrasing her, "There is no situation made worse by having money."

"A wise woman," Pietro said, nodding. "Well, Wittgenstein used the money wisely, commissioning composers to write pieces for orchestra and piano that he could play left-handed. And not just any composers, but the best composers around. Maurice Ravel composed a piano concerto for left hand and orchestra. So did Serge Prokofiev. And young Benjamin Britten. And the great Richard Strauss. Scherchen had liked that piece. Well, if Wittgenstein could play left-handed, why couldn't Cesaro conduct left-handed?"

Pietro paused. He ordered a brandy for each of us. "Lack of knowledge is a wonderful thing," he said. "You and I would have known why Lamberto's idea was doomed to fail. It was a crazy notion, one without a neurological prayer, so to speak. Wittgenstein had lost his right arm, yet his brain remained intact and his musical abilities were unchanged. But Cesaro had a global aphasia. He was a man whose sentences each contained only a single word, sometimes two. A man whose understanding was fragmented. To conceive that such a man with such a brain could conduct was absurd."

Pietro had made his point. The brandies arrived. We each took a sip.

Pietro continued his story: "'Conduct,' Giovanni reiterated. And then he compounded his sin by giving a second command. 'With your left hand. Conduct with your left hand.'"

A two-step command which involved a right-left judgment— no one who understood global aphasia would ever give such a command. It was guaranteed to produce failure and frustration. But Giovanni Lamberto had no such understanding of the nervous system: he was a failed musician and retired second-rate radiologist.

"'Conduct with your left hand,' he ordered his old friend sternly as if he were Maestro Scherchen giving instructions in a conducting class. And, as he gave this command, Giovanni played the first few notes and looked up from the keyboard. It was as if he were the entire orchestra waiting to be led, to be conducted.

"And conduct is exactly what Cesaro Rota did. With his left hand. Making all of the necessary gestures: controlling the *tempi,* the volume, the phrasing. The entire performance. Could Giovanni have played it like that, he too would have been offered a job as an assistant conductor!"

Pietro looked at his watch. It was time to get me back to the hotel.

"But what happened?"

"You saw what happened," he said.

"I need more details," I complained.

"As a neurologist? Or as an author?"

I didn't even bother to protest. I had fallen into the trap. "For me there is no longer a difference. There once was, but no longer."

"Tomorrow I can introduce you to Maestro Rota."

"My flight is at eleven. I have to be at the airport before nine."

"There are direct flights to Chicago every day of the week."

It didn't take much convincing.

That night, as I tried to get some sleep, every article that I'd ever read concerning the loss of musical abilities spun through my mind, as did every case report of musicians with aphasia. Famous musicians like Ravel. Amateur musicians. The issue at hand was not whether musical ability was lost because of a circumscribed lesion in the brain, but rather, what happened to musical talents when there was a lesion that caused a different circumscribed deficit? How circumscribed was circumscribed?

What was music? What was its neurological substrate? How did the brain acquire musical abilities? Not just where, but how and where? Could the neural pathways for music be yet another spandrel?

Music is universal. It's present in all cultures. Melody. Rhythm. The organization of sounds into that planned succession that we recognize as music. That universality seems to indicate that music is not built-in but an acquired function like language. Not acquired like reading. After all, only a few languages ever developed a writing system of their own. But every language group has songs.

I suddenly remembered something my oldest daughter had taught me about music. She had not really meant to teach me anything. She had just made a wisecrack. We had met for lunch as we often do, at a Middle Eastern restaurant. She loves Middle Eastern cuisine. As soon as we walked in she said, "Thank God, they are playing the song."

"The song?" I had wondered aloud.

"The song. There is only one Arabic song, and they all play it. That is how you know the food will be authentic. They play the song."

She knew that there was not just one song, but to ears that

had spent their entire lives listening only to Western music, the entire range of traditional Arabic songs all sounded the same. That's when I suddenly understood that the same basic rules for acquiring language also apply to music. If you grow up in Damascus, you will learn all those songs, that style of music, and they will not sound alike at all. But Western classical symphonic music will all sound the same to you. It's Beethoven writing the same piece over and over again. Sometimes under the name of Haydn, or Mozart, or Brahms, or Mahler. The orchestra is playing "the symphony" again. A learning experience in which environment is pivotal. Listen to Arabic, and you learn Arabic. Grow up listening to English, and you learn English. Grow up with Frank Sinatra and Bing Crosby, and you know the difference between them and the Rolling Stones and The Grateful Dead. Grow up in Baghdad, and you learn to recognize even the most subtle variations of "the song."

As is the case with language, there is a definitive window of opportunity for learning music.

This took care of how (and when) music is learned but left "where" unanswered. And also the question as to whether the brain structures for expressing or performing music might be a spandrel.

Vissarion Shebalin had also become aphasic. He had been an outstanding twentieth-century Russian composer, a contemporary of Prokofiev and Dmitri Shostokovich. How good a composer had he been? It was hard to know for certain. More than just a composer, he'd also been an outspoken critic of the Soviet government. That was not a healthy thing for a composer living in Russia under Stalin. You could lose a lot more than merely your right hand.

Shebalin had had a left-hemisphere stroke resulting in an aphasia and had suffered a severe loss of *expressive* language—that is, he could no longer speak—but his receptive speech was still good and he could understand language. Shebalin had a Broca's aphasia yet continued to compose some of his best works; his receptive and expressive musicianship had both been spared. Those skills were controlled by his untouched right brain.

But what of Rota with his *global* aphasia?

And how much of music depended on a spandrel? Certainly reading musical notation had to involve one as much as reading the written word did. And so did writing down his notations. Both activities have to be taught and learned through formal teaching, as do reading and writing. But all the sets of musical rules, even though they vary from culture to culture, like language, seem to be equally easy to acquire. Everybody learns the music to which they are exposed, if you start at the right age. But not composition—not the creation of a new melody, and certainly not the ability to compose a symphony with orchestration and harmony. Those aspects of musical composition have to be, at least in part, a form of spandrelization. Composing Beethoven's Ninth was nothing less than an absolute triumph of Stephen Jay Gould's spandrel concept. That means that some of the same rules that apply to speech apply to music—The same three basic rules for acquisition—*exposure, exposure, exposure!*

But beyond those basic skills there is a more complex musical experience that requires teaching, that depends entirely on spandrels. And those spandrels have their own series of defined rules. Different rules in different cultures. All are controlled by three basic principles: *exposure, exposure, exposure!*

I finally fell asleep humming the Jean Sibelius's *Violin Concerto,* the revised version as played by Jascha Heifetz. There might be better performances, but it was the one I grew up with.

The next morning before Maestro Rota and I met, I was given the opportunity talk to his therapist. Rota's stroke had occurred almost three years earlier. They showed me the CT scan. The stroke had not been a small localized area of injury. It had been a massive stroke, involving most of the left hemisphere, the entire area supplied by the left middle cerebral artery, which supplies most of the middle two thirds of the left hemisphere. Such patients often do very poorly. They make limited recovery and are left with significant language difficulties of a global variety with both expressive and receptive components. Aphasia and hemiplegia. Far worse off than Paul Wittgenstein.

How had Rota been immediately after the stroke?

He'd had severe global aphasia.

Meaning? Whether it was the neurologist in me that needed the details or the writer didn't matter; I needed the facts.

Rota's comprehension had been quite poor. Repetition had been severely impaired. Speech production had been very limited. He underwent eighteen months of the most intensive speech therapy possible, at the end of which he was again formally reevaluated. Rota still had a severe global aphasia. His understanding was still poor. Not as poor as it had been, but still debilitating. His ability to repeat remained severely impaired. And his speech production was very limited. But he could sing.

"Most aphasics can sing," I reminded them. "Could Maestro Rota sing immediately after his stroke?"

They had never formally tested that.

I imagined Giovanni Lamberto singing to him—that the two of them had sung arias from more than just *Cavalleria.*

Could he read?

No.

What of his musical skills? Could he read music?

They were speech specialists, not musicians. Most of them could not read music. How could they be expected to evaluate that?

I watched Maestro Rota walk toward me. He was accompanied by my friend Pietro Baldacci and another man I recognized as the pianist the night before in the Bloch *Concerto Grosso,* who had played the *piano obbligato* part. The extent of Rota's right hemiplegia was obvious. In a hemiplegia, the muscles of the affected side of the body are not all equally involved.

In the usual stroke, those muscles which extend (straighten) the arm are more affected than those that flex (bend) it. Flexion refers to any movement that bends the joint so that the two bones that make up the joint come closer together. As a result of a stroke, the arm is held in a flexed position, bent at the elbow and wrist by the flexor muscle which are no longer opposed by the weakened extensor muscles. In the leg it works just the other way around. It is primarily the flexors that are weakened while the extensors are in control. Extensors keep the bones in a straight line. In a stroke, the leg is stiff. The hip and knee do not bend (flex) in order to lift the foot off the ground and take the next step. Instead, the unbent leg is pulled around, swinging out to the side and then coming back into position, a movement referred to as circumduction. In a stroke, those muscles originally designed to resist gravity in our four-legged predecessors are not as severely weakened as those originally designed for voluntary movement, even such basic voluntary movements as propulsion. For the maestro, the muscles with the least relationship to gravity were those of the right hand. No wonder Maestro Rota had learned to hold his baton in his left hand.

Pietro introduced me to the maestro. We nodded to each other. He could not even use his right hand to shake hands.

Then I was introduced to the other man, Giovanni Lamberto. Lamberto had become a pianist after all. He had become used to playing under Rota's direction. And Rota had become very used to directing him. The maestro said, "Scuss" and circumducted his way to the lavatory, which was fine because it was Lamberto I really needed to talk to.

I turned to him. Would he mind if I asked him some questions about the maestro?

That was why he was there. Pietro had told him I would be asking some questions.

Could the maestro still read music?

No.

Could he learn a new score by listening to it?

No. He only conducted scores that he knew from before the stroke.

Could he sing songs?

Yes.

Arias?

Certainly. He could sing all of the operas of the standard repertoire. All those works he had conducted for so many years.

Verdi. Puccini. Bellini. Donizetti. Their beloved *Cavalleria*. Even the sextet from *Lucia,* all six parts of it.

Clearly his ability to learn new music had been lost. A partial receptive amusica. A Wernicke's amusica, so to speak.

"But he can learn new songs," Giovanni Lamberto said.

I was surprised. What did that mean? That his receptive amusica was not severe? Or that the answer to my question was not a simple one? It meant that music did, and did not, depend on spandrels. That it was more like reading than speech. Learning a symphony required a spandrel. But learning a simple song depended on a far more basic, more inherent mechanism. Our brains had all been designed to have that skill, or they have selected that skill and have in turn selected to maintain it when given the proper exposure. Rota could forget Beethoven but remember simple songs. It was not exactly an epiphany, but it was something I had not previously understood that clearly.

What sort of songs?

Giovanni was not sure what I meant.

I told him about my daughter and "the song."

He understood. "Old songs," he said. "Classic arias."

"Couldn't he just be relearning songs he knew from before?" Always the skeptic. Too often families tell you of the newly gained skills of patients, grasping for evidence of improvement only to have misinterpreted what they think they have observed. Not new songs for old. But old songs rekindled. Was that what the maestro had "learned"?

"No. They are new to him."

"How can you be certain?"

"A new score has just turned up of a minor work by Rossini. Rota had never seen it or heard it. No one had. I sang a song to him. He learned it. We sang it together. He can learn new songs." There was a triumphant tone in his voice, as well there should have been. Giovanni had been responsible for bringing Rota's music back to life.

"Can he conduct it?"

"I don't know. We never tried."

And right on cue, the maestro returned.

I was not certain what to say. He was aphasic. He had a global aphasia. How global? What would he understand? "The Bloch," I said.

He looked at me.

"Bravo," I said. Had I known how to say "beautiful" in Italian it would have helped.

He smiled.

"Kubelik," I said.

He looked at me curiously.

"Bravo Kubelik. Bravo Rota."

His smile broadened.

"Bravo Kubelik, bravissimo Rota."

"No." He shook his head.

He said nothing more. Giovanni sat at the piano. He started to sing the "new" Rossini aria. Cesaro Rota sang with him. Every word, perfectly enunciated and on key.

Giovanni stopped singing and playing in the middle of the aria and went back to the introduction. He played a few bars, stopped again, and said, "Conduct."

The maestro lifted his left hand and held it there motionless. Giovanni started to play, and Rota tried to keep up. But the pianist was leading the conductor; Rota was lost. Giovanni could not go on and began to sing the aria. Rota began to sing with him, and as he sang, he moved his left arm to keep time with the music. It wasn't conducting, but it was far more than he had been able to do before. So this was the way he relearned, by creating a new method, singing to the music.

So much for my question. The spandrel had lost all of its artistic heritage. It was left naked, stripped bare. And yet the dome had not fallen. Music depends on a spandrel, and it doesn't.

As Maestro Rota began to leave the room, I once again said, "Kubelik bravo, Rota bravissimo."

He shook his head and said only two words: "Schick bravo." Then he and Giovanni left. I knew exactly what he meant. George Schick had been the pianist who had played the *piano obbligato* part on the Kubelik's Chicago Symphony Orchestra recording of the Bloch. Schick bravo! Schick had given the performance of his life, his only legendary accomplishment.

And Rota had known and remembered that.

"Why does he use Giovanni as his pianist?"

"He has no choice," Pietro said.

"Loyalty," I assumed.

"Money," I was told. Lamberto underwrote every one of Rota's concerts.

My mother was right. There is no situation made worse by money. Wittgenstein had his father's money, and so he could commission works from Ravel and Strauss and Prokofiev. Rota had Lamberto and the money to supply an orchestra, but Lamberto came along with it. It gave an entirely new meaning to the word *obbligato*.

MUSIC, OF COURSE, is not a single monolithic mental capacity. It is made up of different functions and aspects, each of which has a different neurological basis. Only now have the individual components come under individual scrutiny. Melody does not have the same anatomical structure as rhythm. And timbre has a different substrate. And composition, which brings all such aspects together, is again different. And with all of these aspects there are critical periods, rather precisely defined windows of opportunity, during which one must acquire the necessary skills. Professional musicians have always known about these windows. In order to become a virtuoso, a musician has to start playing and practicing before the age of thirteen. Not at fourteen. Not at fifteen. Before the age of thirteen.

Neuroscientists have proved this by looking at magnetic resonance images (MRIs) of the brain in violinists while they played. This technique shows which neuronal circuits are activated during a specific activity. They found that those fiddlers who start playing early in life activate larger and more complex circuits than those who start later. The brains of those who start at age three or four look no different than those who start at eleven or twelve. After that there's an abrupt change between the ages of twelve and thirteen. Those who don't start fiddling until after thirteen never catch up. The circuits they activate are smaller, less complex, and more restricted. The biggest differences lie in the size of the corpus callosum, the major bridge that connects the two hemispheres. It was larger and thicker in those who start playing at a young enough age. Has it really grown thicker? No. It has been selected to stay as thick as it originally developed in order to allow the coordinated interchange between the motor areas of the two motor cortices. Which is what violin virtuosi need but the rest of us can live without.

Even more intriguing are the few preliminary studies suggesting that in at least some dyslexic individuals, the corpus collosum is narrower than in those who read normally. Not near the front, in the same area that is thicker in violin virtuosi and where movements are coordinated, but farther back, where receptive language and visual input are coordinated. Are they born with small interconnections? Or are they unable somehow to select this spandrel efficiently enough to preserve it?

Personally, my money is on the latter.

The world of music abounds with such tales of windows of opportunity. According to one such legend, Leonard Bernstein's aunt moved out of Boston when little Lenny, who had never played a note of music in his life, was four years old, and she left her piano with Lenny's mother for safekeeping. It was love at first sight. A musical genius if ever there was one. What natural talent. But what would have happened if his aunt had moved a decade later? It would have been too late. Those areas of the brain that would have to have been selected would already have been selected against.

What would have happened if Mozart had been born in the Stone Age, say, before there were pianos?

PART TWO

. . .

The Brain's Soft Spots:
Programmed Cell Death,
Prions, and Pain

8

—

MY LUNCH WITH OLIVER

·

Why That Morning Was Different from
All the Other Mornings

ALL SUPERFICIAL COMPARISONS to the contrary, Oliver Sacks and I are really quite dissimilar. True, we are both neurologists. True, as neurologists, we both have a major and abiding interest in the same disease, Parkinson's disease. And we both write books about our neurological interests that are intended for a far wider audience than our fellow neurologists. It's also true that our writings have to a large extent grown out of our experiences with our patients and our research on those patients. Too, we both read voraciously. And we could both stand to lose thirty pounds, maybe even forty. I more than Oliver.

Prior to 1986, we had never met. We were members of the same professional clubs, yet traveled in far different circles. We also read—and wrote—very different books: He can write about a patient with Parkinson's disease and relate that patient to James Joyce, D. H. Lawrence, Hermann Hesse, and H. G. Wells. I'd be dealing with Dashiell Hammett, Ernest Hemingway, Graham Greene, and Kurt Vonnegut. As much as I hate to admit it, I never got through *Ulysses,* and never even started *Finnegans Wake.* I have it on good authority that no one has ever actually read *Finnegans Wake.*

At times our literary paths have crossed: we've both read T. S. Eliot. Who hasn't? At least who from our generation with any modicum of a literary background, much less with any literary ambitions? Or perhaps on my part I should say "pretensions." But I distrust Eliot's truth because of his anti-Semitism. Does Oliver? We never once bothered to discuss him. Eliot and his gentleman's club variety of anti-Semitism no longer seem that important to me.

What has become of increasing interest to me is the simple fact that Oliver Sacks and I think about neurology in quite different ways, with perspectives that are barely tangential to one another. That is quite a feat for two neurologists who were trained in the same country at almost the same time. He trained in Los Angeles and I in Minnesota and Chicago, both completing our training in the middle sixties. The brain and how it functions is to Oliver a philosophical issue. This connection to philosophy is, quite simply, beyond me. I have never read Nietzsche. Nor Thorstein Veblen. Nor Bertrand Russell. The only twentieth-century philosopher I have ever struggled through was Karl Popper. To me Schopenhauer is nothing more than a name in a couplet by Lorenz Hart, written to be sung by a stripper who was based on Sally Rand, whose mysteries I have read.

Instead, I fashion myself a classically trained neurologist who thinks about the nervous system within a classical neurological framework and looks at the brain from the classic neurological perspective. I see a patient with a hemiplegia, and I immediately think about the deficit, its location within the nervous system. The disability and *its* meaning come first. Oliver is special in that he has never worn these blinders. He thinks about the brain within a framework that not only has different borders; it also includes aspects that were never even considered in my world of neurology. When Oliver sees the hemiplegic patient, his primary concern is with the function of the rest of the brain, how that brain readapts, and what that means to the patient and his life.

I try to ask simple questions, research questions that can be proven wrong. Oliver tries to understand on a far different plane. He asks questions that are far more profound, almost ephemeral. So our questions come from different angles. Hopefully they are complementary angles and approaches. There are scant areas of congruency, a few degrees here and there. Fortunately, over the past thirteen years, we have also become friends. So when I was in New York, I stopped by his office-apartment, which was then on Twelfth Street in Greenwich Village.

There is another similarity, one which could serve as a major bond. I am certain that in grade school, or whatever that is called in London, he too was one of the last to be chosen for any team. Whether for the sport was football or rugby, baseball or basketball, both of us were among that group who were last to be chosen. "It's a problem," Oliver remarked as we both half-stumbled down the short, ill-lit, worn-down stairway leading to the Italian restaurant he'd chosen for our lunch together.

I had seen him briefly at a premiere of *Awakenings,* the film adaptation of his best-selling book, *The Man Who Mistook His Wife for a Hat,* in Washington the previous week, and told him that I would be in New York and perhaps we could get together. He seemed to be enthusiastic. So I suggested lunch and he agreed and here we were.

He stood at the doorway, pursing his lips and stroking his overly full salt-and-pepper beard—an image of him that had suddenly become part of his public persona after being played by the actor Robin Williams in the film. "These all used to be harmless mannerisms, just little habits that I carried out, part of my unconscious personality. Now they all seem to be caricatures of Robin Williams," he said. "I no longer know if I do them unconsciously for the same reasons I always did them, or because I, for some unknown reason, need to copy his impression of me in order to be me."

Probably both were true. Oliver was able to ponder the issue in his own quirky ways, those idiosyncrasies that were part of how his brain, as a functioning entity, had come to terms with the world as he wandered through it. Robin Williams had obviously studied his subject very well.

We sat. The bread and butter arrived. We devoured it.

We talked of many things including the movie. In an early version of the screenplay, the screen writers had converted him into a woman and Robert De Niro had awakened to fall in love with her. That had fortunately been rewritten, and Robin Williams had become Oliver Sacks, mannerisms and all.

"I am not a character," he protested.

"We all are," I consoled him.

We discussed the status of experimental fetal implants for Parkinson's disease, research I track far more assiduously than he does. My research group had been at the forefront in the study of adrenal implants, and we were about to embark on a program of fetal implants. I brought him up to date.

We spoke about our writing schedules. He went to the hospital only three days a week and wrote on the other days. I admitted my jealousy—not of his success—of his free time. Time protected from other responsibilities. Time to write.

And authors. Not James Joyce. Not D. H. Lawrence. Not Dashiell Hammett. (Certainly not Hammett.) Nor Eric Ambler. But we did have common ground, which we easily found: D. M. Thomas. Harry Mulisch. And a variety of poets: W. H. Auden; Constantine Cavafy; Dannie Abse, a Welsh Jewish physician and poet who lived, wrote, and practiced medicine in London, where Oliver had lived until he came to the United States to study neurology in the mid-sixties. Did he know Dannie Abse? Were they friends?

He and Dannie Abse had been friends.

I sat back. I knew a story was coming, could feel it generating. Oliver pursed his lips and stroked his beard. With a piece of bread, he wiped his plate clean.

I followed suit.

"I was ten years old. It was during the War. Let's see," he paused. Another stroke. He picked up another piece of bread and studied it. His plate had already been wiped clean, twice. He put the piece of bread back. "Nineteen forty-four. No, '43. I'm certain it was in '43. We went to one of his poetry readings."

At the age of ten! I was awestruck. What was he doing at a poetry reading? He should have been playing stickball in the street like I was. Like the rest of us ten year olds.

"I went with my mother. It was at the Cosmos Club. In Swiss Cottage. In London."

"He still lives near there," I chipped in. It was nice to be able to add something substantive to the conversation.

"After the reading, we talked. I asked him about one of his poems."

At the age of ten!

"But we haven't spoken since. Not that I can recall. I think you would have to say our friendship has lapsed."

Lapsed. Nineteen forty-three. Almost fifty years. Age ten!

"You do know Dannie?" he asked me.

"Yes."

"How did you meet?"

"I wrote him a fan letter."

An approving nod.

"That was two years ago. I had never read any of his poems before then."

A disapproving stroke. "Tsk . . ."

"I was thumbing through the book. I saw his poem about Ezra Pound. The concept angered me. I was sure it would be another lame intellectual defense of Pound. Pound the poet. Pound the genius. As if poetry forgave all sins. To say nothing of genius. How could anyone write a poem in honor of that anti-Semitic SOB? I read it, seething through every pore. But it was not an apology. It was an attack, an absolute condemnation of both Pound and his defenders. It was written in a poetic language that defied any attempt at contradiction. Abse had said everything I felt, everything that had to be said, far more articulately than I ever could. About both Pound and his defenders, who came across almost as co-conspirators."

"Abse is a poet," Oliver reminded me.

"And I became a fan of his. I sat down that night and sent him a fan letter. It was the first fan letter I had written since I was eleven and sent one to Jacob Nelson Fox."

"I'm not familiar with his works."

Fox was the second baseman of my beloved White Sox whose games I suffered through while Oliver went to readings by Dannie Abse. There was no reason why he should ever have heard of Nellie Fox. No reason at all. "He only appeals to Chicagoans," I said. "Of local interest only."

An approving nod while twisting of a strand of his beard.

I changed the subject. I hadn't come there to discuss Nellie Fox. I had a story I wanted to tell Oliver. He had written the story of his broken leg and how that had changed his perspective on disease and patients. I too had a story. It was about a broken toe. It had not changed my view of life, only one aspect of how the brain worked. He had gone from loss to perspective, to personality, to philosophy of life. I had gone from deficit to mechanism of how that deficit changes the physiology of the brain. After all, it was my story, not his. I told it to him in as much detail as I could.

It had started at six forty-five that very morning. I was right on schedule. I'd been up writing and sipping coffee for close to an hour and a half. Now all I had to do was complete my shower, shave, get dressed, and I'd be out of the house by seven on my way to the hospital for my day's "real" work. After rinsing the shampoo out of what little hair I have left, I turned off the water, opened the shower door, reached for the towel, began to dry myself off, stepped out of the shower stall with my left foot as I had done an uncountable number of times, swinging my right foot over the ledge, and missed. Pain. Excruciating pain exploded in my head! My right great toe had not made it over the ledge. I had somehow managed to smash it into that inch-and-a-half-high ledge. The toe that I had just broken a few weeks earlier, that was just beginning to heal.

Why this morning? Why not some other morning? Why did I have to be so damn clumsy when I already had a broken toe? Why? The pain was beyond belief. It flooded my brain. My entire foot was aflame. My whole leg. On all other mornings without my looking, my foot glided safely above that ledge like a good tennis shot. Why, when I knew I had a broken toe? I was not that much of a klutz.

But suddenly it made sense: I had been suckered by my own brain, by some primitive defensive system of my brain that I'd never even known of—me, a neurologist. It wasn't my fault. It was my brain's fault. And as my toe throbbed and ached away, I recognized that I had learned something about the brain, something that explains events that happen to everyone usually more than once but that remains unmentioned in any neurology book or journal I have ever read. Eureka. Land ho!

Oliver frowned. I had his complete attention. No bread. No beard. No bread sticks even.

Three weeks earlier I had broken my toe while tripping coming down the stairs at 5:00 A.M., a result of not turning on any lights and of my own clumsiness. As I said, it was a fracture of the big toe on my right foot, right at the joint. I diligently taped it to my second toe and went on with my business. This included a whirlwind professional trip to Europe which featured four different countries in six days and a lot of walking through airports and various other places. None of which helped the healing process. But now, three weeks after my original fall, my toe was finally improving. I could now stand without discomfort and walk without pain.

And then calamity struck. Why? Because, simply put, the toe had not healed completely. And if I had been paying attention, I would have known that I was still experiencing some pain. My supposed painlessness had been an illusion, a conjurer's trick perpetrated by my own brain. True, I was walking without consciously feeling any pain. But there is a difference between absence of consciously perceived pain and freedom from pain. A big difference. Just ask my toe.

If a tree falls in a forest and there is no one there to hear it, does that crash make any noise? You know darn well that it does. Oliver nodded. A simple nod. Nothing more.

And does a broken toe cause pain each time you slam your weight down on it even though you feel no pain? Hell yes!

Then why didn't I feel the pain as I stood in the shower that fateful morning? We call it tolerance. You put on a tie in the morning, the collar feels tight around your neck. Half an hour later you can't feel it at all. Or you put on your bra and the straps pull on your shoulders. Not for long. In a few minutes you don't feel them. It's as if they have disappeared. The straps, I mean.

Why?

Tolerance. The steady ongoing stimulus of the collar or the bra no longer registers consciously in the brain. Why not? Teleologically, it is because that unchanging stimulus serves no purpose. If it were otherwise, our consciousness would consist entirely of myriads of inputs of unchanging messages: the feel of our shoes, our socks, our pants, our underwear. Without thinking about it, do you feel your underwear? No. Yet your underpants are touching your skin and stimulating the nerve endings there. As a result of that stimulation those nerves send messages toward the spinal cord and then up to the brain. For you to consciously feel your underpants, all you have to do is think about them on your buttocks.

But why don't you feel them most of the time? Because your brain adapts to a fixed, steady input and no longer "feels" it, or to use the preferred neurological term, your brain develops tolerance. The stimulus is the same, but the response is gone.

Pain is just one of a number of primary sensations the nerves send up to the brain. Others include position sense (where a part of the body is in space), touch, temperature (hot or cold), and pressure. Each of these is subject to tolerance. That morning, when I first got in the shower, the hot water quickly shocked my body. But by the time I got to the second chorus of *Our Love Is Here to Stay* (I always sing show tunes in the shower), I had to make the water hotter. Tolerance. The same tolerance that keeps me from being plagued all day long by that collar around my neck. And causes me to feel for my watch to know it's still there. And my wedding band.

The pain from my broken toe had obviously not gone away because my toe couldn't heal that quickly. Not in a fifty year old. All I had to do was look at it to know it had a ways to go. It was still enlarged. The soft tissue was swollen. The bone itself was thickening as new bone was being laid down. And if I tried to move the toe, I realized how far it was from being back to normal. The joint was half frozen, hardly capable of any movement at all. And any movement I forced it to make was painful. Not uncomfortable, but downright painful. Or when I palpated it with a firm push, the toe was tender.

If pushing down with my thumb caused pain, if standing up

on my foot the first thing in the morning caused pain, didn't standing on my right foot in order to wash my left foot, and thereby pushing down hard with my right big toe, make my right toe and its sensory receptors, the raw nerve endings, feel and produce pain?

It did. But I felt no pain as I went right on singing. Bernstein. Then Sondheim. *Ladies Who Lunch,* I think. I always loved that one line in which he rhymes "piece of Mahler's" with "thousand dollars." Lorenz Hart himself could never have done it any better than that.

Each of us knows precisely where each part of our body is at every moment. That is because each part of the body—or more precisely the nerves of each part of the body, especially the nerves of the joints—sends messages informing the brain of its location and position in space. The brain keeps track of all of this information unconsciously, automatically, reflexively. We do not become consciously aware of these facts unless we move a part of our body or ask ourselves a conscious question— where is my left leg positioned right now? let's say. If there is a preconscious part of our knowledge that can effortlessly be called into consciousness, this is it. When you move an arm, you become conscious of its position, only to be unconscious of it seconds later when your attention is turned elsewhere. Our brain's ability to divert these sensations out of conscious awareness is an example of the brain's ability to adapt: we no longer consciously need to know our position at all times.

I looked at Oliver. Did he want these details? He did. I was engaging all of his faculties. All of his vocations: Neurology, writing, philosophy—all three. He was the perfect audience. I went on.

Each morning I get into my car, I duck my head and miss the top of the door frame. I've never hit my head on the door frame of my own car. Not once. Have you? I don't think about hitting it. I don't worry about it, just like I don't think about smashing my foot into the ledge around my shower stall. But if I put on a hat, that hat hits the door frame as I slide into my seat unless I direct my attention to it.

Or I'm writing at my desk, completely absorbed. I reach up

with my left hand and run my fingers through what remains of my hair. Once. Twice. Three times. I know what I want to write. I start writing. I write all first drafts longhand. Without thinking, I put my left elbow back down on the desk. Softly. A perfect landing. I never hit the "crazy bone," the nerve running along my elbow.

Why not?

Because I know where my elbow is in space; or more precisely, my brain knows it. It's called unconscious proprioception, the unconscious knowledge of the position in space of each and every body part: of my elbow as I rest it back on my desk; of my head as I slip into the car; of my right foot as I get out of the shower.

But I smashed my right foot with its broken toe into the ledge. Ergo: I didn't know where my right foot was in space. I didn't know it consciously. Of course, I wasn't supposed to know that fact consciously. I was too busy working my way through some Gershwin tune. But here's the rub: I didn't know it *unconsciously* either. That was where my brain had screwed up.

So why didn't my brain know where the hell my toe was?

The answer was obvious, like all such simple truths. I had not only developed a conscious tolerance to pain, I—or rather my brain—had automatically developed tolerance to unconscious proprioception, to where my right toe and right ankle were in space. My tolerance had spread across two separate borders: from one order of sensation (pain) to another (position sense, or proprioception), and from consciousness to unconsciousness.

This kind of tolerance happens to all of us. Say you're working around the house, hammering away. You hit your thumb with the hammer. Accidents do happen. Your thumb hurts. It kills. Then it feels better, so back to work. More hammering. Within no time, half a dozen hammer blows, you've done it again, smashed the same thumb with the same hammer. Why? Because in dampening the original pain the brain has also dampened its knowledge of where in space that part of the body feeling pain was. Where your thumb was.

And bang, you do it again. Pain. Worse pain than before.

Pain is unique among sensations. It is the most primitive, yet it is an integral part of the nervous system serving to protect the organism. As such, the pain sensation retains many of its primitive features. The nerve endings that feel pain are simply the thin, naked endings of the nerves. Other sensations require specialized receptors. Not pain. Just the nerves themselves are required. Moreover, while all other sensations reach consciousness in the cerebral cortex, the mantle of gray matter and nerve cells on the surface of the brain, conscious awareness of pain is felt deep in the brain, in the thalamus. Our cerebral cortex is by far more developed and sophisticated than in any other species: destroy or anesthetize the cortex, and the conscious recognition of sensation disappears: hearing, vision, position sense, touch, smell, temperature. But not pain. Pain is felt consciously in the thalamus.

Pain recedes in a few minutes. Slowly. Surely.

Thank God for tolerance. You start to work again. And bang. *Disaster.*

The thalamus lacks sophistication. It has no judgment. It dampens pain and as it does, it also blocks position sense, preventing the latter sensation from reaching consciousness and even aborting its unconscious awareness. That is why I hit my thumb that second time and the third time, and why I hit my broken toe on that ledge.

I finally understood why the accident had occurred on that particular morning.

I was finished with my story. We got up to leave. "I know a great place for dessert."

"Lead on," I said.

Perhaps we are not all that different: forty pounds plus one dessert.

9

—

TWO SETS OF BRAINS

·

Something Old, Something "New"

OUR FASCINATION WITH dinosaurs is as old as, well, since humans have been around. We've always been interested with them. *The Lost World* was a major work of fiction and a big moneymaker as a movie long before Michael Crichton or Steven Spielberg made their millions on the project. I was seven years old when I fell under the spell of dinosaurs. The great debate as to why the dinosaurs had all died off had no real answer back then, even though most kids who had seen Walt Disney's *Fantasia* half a dozen times understood it had something to do with the changes in the environment. But then I was seduced away by the far flashier allure of baseball. Like hundreds of boys my age, I became obsessed with such questions as: Which was the greatest baseball team of all time? The '27 Yankees with Babe Ruth, Lou Gehrig, Earle Combs, and Poosh 'Em Up Tony Lazzeri? The '29 Philadelphia Athletics with Jimmy Foxx, Al Simmons, Black Mike Cochrane, and Lefty Grove? The 1919 Chicago

Black Sox with Shoeless Joe Jackson, Eddie Cicotte, Buck Weaver, and Eddie Collins? Walt Disney couldn't dream of more hard-driving questions.

The change in perspective which came to rule the next few years of my life and left an indelible imprint on my mind and soul reached so far that it even affected my all-but-unformed taste in literature. My old poetic allegiance switched to the immortal classic "Casey at the Bat." Before that, my favorite poem, and the first I committed to memory, had been a bit of extended doggerel about dinosaurs which I had read in a book published by the Museum of Natural History:

> You will observe by these remains
> The creature had two sets of brains,
> One in the head, the usual place,
> The other in the spinal base.

In retrospect, this was a most interesting fragment for a seven-year-old future neurologist to repeat *ad nauseam* and then store away deep inside his brain, all but forgotten until decades later, when it sprang forth unprompted.

My prompt was a patient, Rosemary Covington, who wanted to know what was causing the disease that was driving her crazy.

Rosemary Covington—or Mrs. Travis Covington, as she often referred to herself—had traveled from Louisville to Chicago to see me. Her physician there did not know what was wrong with her, but recognized that she had a "movement disorder," a neurological disease characterized by the occurrence of abnormal involuntary movements. This diagnosis did not take too much clinical acumen since one of her two complaints was that the toes of her left foot were moving constantly. Her other was that her foot hurt. She'd had pain in her left leg and foot for years, as well as a bad back, causing her to undergo several operations. She had told her hometown doctor that this pain was different. Her doctor, however, recommended that she go see an expert on movement disorders.

Mrs. Covington was given a choice—New York or Chicago. She opted for Chicago, not because I was a great neurologist, but because it was close to Louisville. She also had a cousin who lived in the Windy City. So Chicago it was, and she duly arrived in my office.

As always, I observed her as she sat in the waiting room. She was very properly dressed and, despite looking at her for several minutes, I could see nothing out of the ordinary. No abnormal movements, which in and of itself was out of the ordinary and eliminated most of the diseases for which patients come to see me. I called her name, and she got up out of the chair without any difficulty and walked into the examining room with an entirely normal gait. More possible diagnoses went out the window. My list was getting very short.

Once I had introduced myself, I asked her what had brought her to Chicago to see me.

"Two problems," she began. "A painful foot and moving toes."

And in saying those two phrases she had handed me the diagnosis. No wonder I had seen nothing. Her toes were moving inside her shoe. "Tell me about the pain," I said.

She did. She'd had back pain off and on for years.

How many?

"Twenty, at least. I was about thirty when my back gave out, and I'm fifty-one now."

I needed more details than that.

"But this pain is different. Very different."

"I'm sure that it is," I agreed. "But still I need to know all about your back problems."

She told me the whole story. She had been working part-time then, with two small children. The younger one had just started school. He was five. So she got a job working at a clothing store, and one day she bent over to lift a very heavy box when all of a sudden this terrible pain hit her smack dab in the middle of her back. She could hardly straighten up and had to go home. The doctor gave her some pain pills and some other pills, muscle relaxants, she thought, and kept her in bed for a couple of weeks. She began to recover and was soon able to go back to work.

I asked her to describe the pain.

It had been terrible. Like a hot knife, starting in the low part of her back and shooting down her leg.

"Which leg?"

"My left leg," she said. "Like a bolt of electricity."

Both the localization of the pain and its character were precisely what I would have predicted. Sharp, shooting pains are the result of direct involvement of a nerve as it travels from the spinal cord to its destination. The best example is your funny bone, which is not a bone at all, but a nerve—the ulnar nerve, which travels over a groove in the bone of your elbow. If you hit that nerve, a sudden electric pain shoots down your arm and into your hand. In that case, the nerve itself is the seat of the pain, and the pain is distributed wherever that nerve normally serves. In the case of the funny bone, pain is distributed along the ulnar nerve.

Mrs. Covington had pain in the distribution of a nerve rooted in her spinal cord which then traveled to her left leg and foot. Nerves leave the spinal cords as roots, one for each level of cord. This is a remnant of the early development of our very early ancestors, long before the dinosaurs. As the original nervous systems evolved, the body was made up of segments. Each segment had its own muscles and nerves and overlying skin. And the nerve of each segment started out as a group of cells within the primitive, segmented spinal cord. Part of us is still organized exactly that way.

Every time I teach neurology to a new group of residents or medical students, I try to make the segmentation of the body clear to them. By the time I meet them, these new doctors or doctors-in-training have all taken anatomy and spent a year dissecting the muscles of the human body and even more years looking at human bodies or well-made sculptural representations of the human form.

I ask them a simple question: "Phylogenetically speaking, what is the oldest and most primitive muscle in the human body?" And they look at me as if I had asked them what part of the body was descended from aliens.

I then talk about the early organization of the body into segments. Each section possesses its own nerves, muscles, and adjacent segmental muscles, which then fuse together with a thick band of connective tissue. Today our muscles are far more integrated. They attach not to each other but to the bones that they move. In our primitive ancestors, these muscles were more like those of snakes—resulting in undulating rather than independent muscle movement. Except that ours were even more primitive. Now we have only one truly segmental muscle, one that is divided into segments, each of which derives its innervation from a single spinal cord level, and each of which is connected to its adjacent segment by a thick, fibrous band called an aponeurosis. Sometimes at this point one of my students gets the answer to my question. Sometimes I have to go on talking a while longer.

Then—bingo! An answer: the rectus abdominis, the oldest muscle of them all. It is the long, flat muscle that makes up the front of the abdominal wall. When it's well developed, you can see how the individual segments are separated from each other by nonmuscular connective tissue. This is known in certain circles as "washboard abs." No matter what you do to other muscles in the body, this delineation can never be seen because no other muscle retains its original segmental organization.

There is even a neurological sign based on this segmental organization. If you are lying on your back and lift your head, the entire rectus abdominis contracts. Your abdominal wall becomes uniformly tight. This is a form of isometric exercise in which your umbilicus, or navel if you prefer, stays in place because both the segments above it and below it are contracting. However, if there is a problem within the spinal cord causing messages to get to the upper segments of the cord but not the lower segments, what happens is very different. You lift your head. The upper segments contract; they got the message. The lower segments don't; they didn't get the message. Your umbilicus moves, pulled upward. This is Beevor's sign, named after Charles Beevor, an obscure neurologist who first described it as a sign that there might be disease in the spinal cord between the upper and lower abdominal segments. This simple yet easily forgotten fact is all based on the primitive, segmental organization of this one muscle. Phylogeny at the bedside.

Mrs. Covington's original pain was due to a radiculopathy: one of the lumbosacral nerve roots that innervated the muscles and skin of her left leg was the site of her pain. The cause was a ruptured disc that was pressing on the nerve root. Her back pain was due to reflex spasms of the muscles adjacent to the spine. Bed rest, anti-inflammatory drugs, muscle relaxants, and the elixir of time helped. She got better and was able to go back to work. But the pattern was set. Every year or so she had another attack, and each was worse than the previous one until she got tired of it all and went to see an orthopedic surgeon.

"When was that?"

"Let me see. My son was ten, I'd been putting up with this for five years."

"Was the surgery a success?"

"Sort of. The pain in the leg stopped."

That meant that the pressure caused by the herniated disc pushing on the nerve route went away.

"But I still had a lot of trouble with my back."

The structural abnormalities of her back that had led to her having herniated a disc in the first place were still there, and they still caused low back pain.

"And the pain in my leg only stopped for a couple of years," she sighed.

The rest of her tale was all too familiar: further attacks of nerve root pain, always radiating into the left leg. Another back surgery for another disc. Limited relief for a year or two. More radiculopathy. A third operation involving fusion of vertebrae. That had been only three years ago. The leg pain had gone away, but the back was a different story. It had improved, but she always knew it was there. Then about a year after her last operation—and I knew by the way she said it, she really meant "last," not just latest—she started having pain in her left foot. "A new type of pain," she explained.

"New in what way?"

"Every way."

I didn't have to press her for details. She was happy to supply them. No one else had ever wanted to know what it was that made this pain different from all of the others she'd previously had in her left foot. The character was different, she said. These pains did not shoot down her leg or shoot at all. Nor were the feelings similar to an electric shock. It was a cross between a burning pain and what it feels like when a piece of metal hits a tender spot on one of your teeth. Sheer agony. And constant. It's always there and always the same.

"And where is it exactly?"

"I wish I could tell you. It's not like my old pain. That started in my back and shot down my leg. I could point to every inch of skin it crossed. Even after it had gone away." True radicular pain follows the path of a nerve root as it travels down the leg. "This pain is never in my back. Or my thigh or my leg. Just my foot. And it's all over my foot and nowhere at the same time. Do you know what I mean?"

The strange part is that I did. It was clearly not a well-localized pain caused by an injury to a specific nerve that supplied a specific location, but a poorly localized pain due to abnormal functioning of a pool of relay neurons somewhere above the offending nerves.

"Then my toes started to move."

I asked her to take off her shoes. The toes of her left foot were continuously moving. It was not a tremor. We define a tremor as a rhythmic back-and-forth movement at any one joint. All her toes were moving in succession, one at a time: up, then down, then back in place as the next toe began to move. It was not dystonia: if it had been, the toes would have remained for a time in the abnormal position. These toes just moved. They formed no postures at all. It was not a type of chorea, as there was nothing at all random about these movements; they continually followed the same pattern, which resembled a crowd at a tennis game following a match: heads moving in a wave, time after time for as long as I watched. Even when she stood up and put weight on her foot, and even when she walked, the wave kept right on going. It was enough to drive a person crazy. Which is just what it was doing to Rosemary Covington.

"Tell me then, is that all?" I asked her.

"Isn't that enough? It's really gotten to me. I've got a painful foot and moving toes. So what do I have?"

"Painful-foot-and-moving-toe syndrome," I said rather sheepishly.

"Don't make fun of me," she said.

But that is indeed what it's called. The name is purely descriptive, and the symptoms are the same in every patient: One foot has pain that, like hers, is hard to describe exactly and even harder to locate precisely. And all of the patients have a past history of prior painful conditions, most of them bad backs with multiple episodes of leg pain and numerous operations. Then at some point, pain and the movements start, so it's called painful-foot-and-moving-toe syndrome. This was one of those times when I really wished there was an eponym.

I told her all that I knew about the condition, which wasn't very much. The recurrent pain seemed to have done something to the pain relays somewhere in her brain or spinal cord. Pain that is poorly localized and of a peculiar quality like no other pain almost invariably has its source not in the nerves but within the central nervous system. The nerves are designed to carry specific messages to the brain, and the brain is trained to recognize and respond to those messages. Part of that response in humans is to localize (identify the source of) the sensation, and to describe it always in terms of some other sensation that has previously been received, classified, and given a name. Ironically, because she couldn't localize the pain, I could tell her that her condition in some way related to the mechanism of her previous painful conditions.

So far not too satisfying. It was about to get worse.

"What about the movements?" she wanted to know.

Those I didn't understand at all.

"But you're supposed to be an expert on movement disorders. That's why I came all the way to Chicago. Maybe I should have gone to New York."

I almost voiced my agreement.

I knew a lot about all of the other movements I have seen. I could tell her all there was to know about tremors. But there was nothing else to tell her about painful-foot-and-toe syndrome.

"So you can't explain the movements?"

"No."

"Did you know how to treat them?"

I didn't.

"Do you know how to treat the pain, at least?"

I didn't.

"So all you know is that the back pain seems to have started this."

She was right about that.

"Does it only happen in the feet?"

"Yes."

"Why is that? Lots of people have neck pain. Almost as many as have low back pain."

She was right about that too.

"And in most of them it shoots into one arm or the other. Or both."

Right again.

"So why don't they get painful hands and moving fingers?"

Something about her question—what is it about moving toes and painful feet?—snapped the long-buried poem back into my brain: "Observe by these remains," I paraphrased, "the creature had two sets of brains: one in the head, the usual place, and the other in the spinal base."

That explained everything. And because the patient had asked the right question, I had the right answer. It had to do with the organization of the nervous system. Sometime after the segmental organization evolved in primitive animals, the nervous system started to develop connections between spinal segments so that multiple segments could act in unison—hence

Beevor's sign. Eventually this organization became located at the two ends of the cord: one in the head, the usual place; the other at the base of the spine. In early vertebrates both were important. For example, both were important in dinosaurs. Both are still important in chickens, which can run around after their heads have been cut off because the brain in the spinal base directs simple reflex walking. And it explained why I had never seen movements like hers in any of the usual movement disorders, which are all diseases of the brain. Her movements had nothing to do with the brain. They originated in the spinal cord.

Man still has a trace of this second "brain" in the base of the spinal cord. Newborn infants have the walking reflex, in which their legs make walking movements. This reflex originates in the lower spinal cord. But infants have no such reflexes in the upper extremities, showing how we, as humans, have undergone progressive cephalization. That is to say, the more advanced movements which infants reveal as they grow are evidence of the tendency for humans and other animals to have evolved so that important organs are located near the brain rather than the spinal base. Our hands and arms are completely controlled by the brain. Nothing is left of any spinal control center for the upper body. Hence no moving fingers.

Trying to control my own excitement and flush of discovery, I tried to explain this as carefully as I could. Her pain was emanating from similar reverberating sets of interrelated neurons in the base of her spinal cord. No wonder the brain could not locate, classify, or describe it. It was not the type of pain the brain was used to interpreting.

"So this is all because I have the spinal cord of a dinosaur."

That was not what I had said, but she had grasped the general idea.

She was far less excited than I was. She looked at me carefully and asked another question.

"Now that you understand my . . ." She still couldn't bring herself to say it.

"Painful-foot-and-moving-toe syndrome."

"Yes. Well, will you now be able to treat my pain?"

"No," I conceded.

"Or the movements?"

"Probably not."

"Then what good does your knowledge do me?"

I shrugged my shoulders.

"Or you for that matter?" Then she added, "Unless of course you write an article about it. That is what you doctors do, isn't it?"

"Yes, it is what we do," I admitted.

"You do that," she said, "and maybe some other, smart doctor will read it and figure out how to treat me. Or somebody else with . . . this same disease."

She understood the entire process of scientific progress.

"Most likely after I'm dead and gone," she quipped.

Then I remembered the last line of the poem. It just sort of popped out:

Extinct ten million years, at least.

10

—

ANTICIPATION

·

Unto the Third Generation and Beyond

THE FIRST TIME I encountered a member of the Hamilton family, we were not properly introduced. She knew my name, but I had no idea who she was. Little did I anticipate that I would become a recurring theme in her family's lives, and they in mine, for the next three generations.

On a Sunday afternoon at a meeting of the Chicago chapter of the National Huntington's Disease Association, I was the medical advisor of the group and was asked to answer questions about Huntington's disease. This was in 1969 or 1970, and what we knew then about the science of the disease was not very helpful. I tried to describe the process of progressive nerve cell death in Huntington's and how that only occurs to certain sets of brain cells. I even talked a little about my own research, which focused on the mechanisms that might produce the abnormal movements characterizing the disease in most patients. Huntington's chorea, as I had always called the disease. It was one of the first disorders to be named after an American physician. It was also one of the diseases that put American medicine in general, and American neurology in particular on the map of world medicine, a great step forward at the time.

During my talk, a woman with a determined look raised her hand.

"Why," she challenged me, "are my son's brain cells dying?"

"I have no idea," I answered, perhaps a bit too glibly. While I had begun to treat the effects of the disease, I had not yet gotten a handle on the cause.

"Some expert," the woman snorted not quite under her breath.

"It's not just Huntington's," I said, attempting to recover some ground. "I don't really know why brain cells die in any neurological disease. No one does." It was a rather feeble excuse, but true.

"And when do you anticipate knowing why the cells die off?"

"Not in the immediate future," I said sheepishly, presenting my most optimistic guess.

After the meeting was over, the woman approached me and told me that I was going to become her son's doctor. She was going to bring him to Chicago from southern Illinois. I might not know all that I should, but at least I seemed to be interested in her son's disease, which was more than she could say for his other doctors.

It was not an appointment that I anticipated with any relish. This woman would not be easy to placate with the usual sophomoric answers and platitudes that most patients want when dealing with the insoluble, unanswerable, and untreatable. She wanted answers, not euphemisms.

She seemed to be brimming with anticipation as she walked away. Anticipation—a word with many meanings. Its scientific definition reverberated through my mind. The connotation derives from the clinical observations made of patients with Huntington's disease and dates back to the "antediluvian" era.

This was before we knew that DNA was the basis of the genetic code, back when science thought it knew for certain exactly how many chromosomes made up the human genome (an as yet uninvented term) but was off by one pair.

In the study of genetic diseases, "anticipation" describes a situation in which a dominantly inherited disease makes its clinical appearance earlier and earlier in successive generations. Beyond this definition, the word has other subtle connotations in the scientific world.

The initial model of anticipation was Huntington's chorea, named after George Huntington, who described the disorder that transformed him from an obscure, small-town general physician into an eponym. Thirty years later, when George Mendel's work on inheritance of traits was translated into English, the disorder Huntington had described quickly became the perfect example of a pure, Mendelian-dominant disease with complete penetrance, that is, a disorder in which every individual with the gene develops the disease and no one without the disease can transmit it to their offspring. In such a case, there are no true carriers.

How could a local physician from rural East Hampton, Long Island, make an observation which had escaped several generations of specialists in neurological diseases? As Huntington himself explained, he was not the first Huntington to have observed these patients. His father, and before that his grandfather, had been general practitioners in the same location and had observed and recorded the same families with the same diseases. Thus, George Huntington had at his fingertips the clinical observations of three generations of physicians who treated the same families with the disease. He noted three marked peculiarities in Huntington's chorea: it is hereditary; along with its other symptoms, it brings a tendency to insanity and suicide; and it manifests itself as a grave disease only in adult life.

The word "chorea" was derived from the Greek word for dance and was used to describe the hysterical plagues of dancing that swept over Western Europe at the end of the fifteenth and beginning of the sixteenth century. At the time Huntington used the word, "chorea" described any disorder causing a peculiar dancelike gait. We no longer use the word in that way. "Chorea" has become the official definition of a class of abnormal movements that characterize not only the disease described by Huntington but also many other diseases with similar movements. These movements are single, individual muscle jerks or contractions that are random in location, direction, amplitude, and frequency. It is their lack of regularity that, in the eye of the trained neurological observer, characterizes and differentiates them from the other classes of movements.

Once Mendel's work became known, Huntington's chorea became a hot subject for medical research. Huntington himself felt that the disease followed some fixed laws, and his initial description revealed that the disease followed Mendel's laws perfectly: it was dominant, and there was complete penetrance with no asymptomatic carriers. Huntington was wrong about a few of his observations. He suggested that it was more common among men than women, which isn't true. And, of course, it is not restricted to a few families in East Hampton, Long Island, but has a worldwide distribution, having been described in virtually every known population group. The families that Huntington described were traced back to their arrival in the New World in 1630 and then to their family origins in Surrey, England.

One other conclusion grew out of George Huntington's studies. He understood that the disease started earlier in each generation. With this observation, the scientific concept of anticipation was born.

Very soon, however, what clinicians had called anticipation was shown to be nothing more than an artifact of prejudiced observation. Once physicians knew it was a dominant disease with complete penetrance and went out to look for families with the disease, they found patients in the earliest stages of the disorder, family members who felt normal. Had these patients not been examined, the onset of their disease would never have been diagnosed so early. And of course the date of onset in the parents was based on medical investigation of living descendants who remembered their parents and then could give an early date of the first appearance of the initial symptoms—the first abnormal muscle jerk, for example. But who remembered the grandparents and their abnormalities, much less the great-grandparents? No, anticipation was a bias and had no scientific basis whatsoever. That was what I was taught, along with the statistical models demonstrating that anticipation could not be real. The idea of "anticipation" was quickly dismissed.

Huntington also used another word in a way that is no longer acceptable: the so-called insanity that he attributed to patients with hereditary chorea was usually not a true insanity, but a type of slowly progressive dementia.

FOUR WEEKS AFTER my lecture, the woman I had met arrived in my office along with her son. This time she introduced herself—Mrs. Janet Swisher Hamilton and her son Bill Hamilton, who was thirty-five years old at the time.

There was no mystery to why Mrs. Hamilton knew exactly what was wrong with him. Her husband, Dave, who had been diagnosed as having Huntington's chorea, had died chained to a bed in a state hospital. Dave's mother, Bill's paternal grandmother, had also died in the same state hospital, chained to a different bed. Autopsies on the grandmother and the father had both revealed the pattern of loss of brain cells that was diagnostic of Huntington's chorea.

Mrs. Hamilton knew precisely what was wrong with her son, yet emotionally she still needed an expert to confirm it and to then attend to her family's illness.

All the while they were in my office, Bill Hamilton said nothing. But his mannerisms confirmed everything his mother said. He sat there displaying a variety of random, purposeless movements. A grimace. A toss of his head. A jerk of his right shoulder, his left hand.

As I put together the family tree, I asked the usual question: what had she observed in her son?

The same things she had seen in her husband.

Which were?

"They both changed."

"In what ways?"

"They were both stubborn men. Dave was stubborn from the day I met him. I thought he was the most stubborn man I ever met. And the most determined. He was determined to marry me and, well, I finally gave in. Over the years he just got more and more stubborn and set in his ways. Then he started to have trouble controlling his temper. The twists and jerks began after that."

"How old was he when it all started?" I asked. The average age of onset of Huntington's chorea in the United States is about thirty-eight.

"When I really knew he was sick?"

"No, when you think that he first showed any changes at all?"

"That's a different question," she observed correctly.

"Yes," I agreed. "That would be the age when the disease really started."

She nodded her agreement and thought carefully before answering, "Thirty-seven or maybe thirty-eight."

Average age of onset, I thought as I recorded that fact in her son's chart. "And Bill?" I asked looking at her son, who grimaced again. It wasn't exactly the same grimace as the first one. The first had been almost a sardonic smile. This was a sneer, more pronounced on the right side of his face. Then another jerk of the shoulder, far more forceful, followed by an intentional movement of his hand to brush back his hair which made it seem that the jerk of his shoulder had been part of some planned, intentional movement, instead of the other way around. Chorea. The diagnosis was self-evident.

"I noticed the same things," she informed me for the second time. "He got more stubborn, more irritable, and then began to twitch."

"When?"

"About six years ago?"

"About when he was thirty-one?"

"I guess so, but of course I knew what to watch for and I was

watching him like a hawk. Him being so much like his father."
The well-known fallacy of anticipation: observer bias, not scientific fact. Even the families knew it.

I looked over to Bill. This time the shoulder shrug was neither a choreatic jerk, nor a masking movement. It was more like a subtle tic, a gesture of nervousness, a movement that most people would see but yet not really notice, a part of the normal repertoire of life.

Mrs. Hamilton understood that anticipation was in the eye of the beholder. She also knew that her son had Huntington's chorea. She had come to me to confirm her diagnosis, which I did. Each choreatic movement is a single, isolated muscle contraction—a short, uncoordinated jerk. These jerks can be located in virtually any muscle of the body causing that part of the body to move in an unpredictable way. The successive occurrence of two or more such jerks can result in more complicated movements, while the superimposition of purposeful movements can result in far more complex movements, such as the one I had observed when Bill Hamilton attempted to link the violent jerk of his shoulder into a voluntary action of smoothing his hair. These choreatic movements are the most characteristic manifestations of Huntington's chorea. George Huntington had seen them in his patients. I had seen them in mine. Mrs. Hamilton had seen them in her husband and her son. They had been the same in all of these patients.

Mrs. Hamilton also wanted to see if there was anything I had to offer her son.

"To control the movements?" I asked. We had some medicines that could do that.

"No, to control him."

I wasn't sure what she meant.

"At times he just plain loses control of himself," she said.

"Physically?"

"Yes."

"Sort of like rages?" I inquired inarticulately.

"Sort of," she concurred.

I put him on a moderate dose of an antipsychotic drug.

Developed to alter the abnormal behavior in schizophrenic patients, the class of tranquilizers known as neuroleptics often also helped patients with Huntington's chorea to control their choreatic movements and at times their abnormal behaviors as well. I never saw Bill Hamilton again.

I knew the Hamiltons carried the gene for Huntington's, but I had no idea what that really meant. This was over twenty years ago. At that time, we could not detect genes. We could only ask questions so as to put together information about the mechanisms whereby the brain produced chorea.

But why do patients with Huntington's disease develop chorea rather than some other form of abnormal movement? Why any movement at all? George Huntington never ventured a guess. It was not an issue that concerned him. But I was a neurologist with a special interest in abnormal movements. My consultation with Mrs. Hamilton and her son Bill had kicked off in me an obsessive need to know more. For a year afterward I read everything I could find on the subject but found nothing enlightening. As far as I could tell, the precise mechanism that produced chorea was not very clear to anyone. Several well-respected authorities attributed the movements to a loss of the normal filtering mechanism in the corpus striatum, that collection of neurons, deep within each cerebral hemisphere, that comprised the motor control area most affected in Huntington's chorea. But this theory was all but impossible to interpret in terms of biochemistry or cellular physiology, which made it impossible to test, and an untestable hypothesis serves no purpose.

It was that interface between biochemistry and physiology that particularly intrigued me. If there was a chemical or physiological change in the brain, how did that produce chorea? The discovery of decreased levels of dopamine in the brains of patients with Parkinson's disease had led to the first successful medical treatment of it. Patients could be treated with levodopa which replaced the missing dopamine. This changed the lives of patients with PD. Could the same thing be done for Huntington's chorea?

Like just about everyone else in the field, I agreed with the theory that the abnormal movements that characterized Huntington's chorea were related to the cellular destruction observed within the striatum. Whatever the as-yet-unknown gene did, it did to the neurons in that part of the brain. The neurons of the striatum die off progressively as a result of the genetic defect. All theories related the appearance of chorea to loss of function of these dead neurons. I began to wonder if there might not be another possibility. Chorea might not mean that the cells were dead. Perhaps it could also result from the abnormal functioning of diseased, but still surviving, cells of the striatum.

This was far more than a subtle difference, one that neurologists are all too familiar with. Paralysis in a stroke is due to loss of functioning of dead cells. Once the natural recovery is over, there is little we can do to change this. Seizures, on the other hand, are due to altered functioning of living neurons, and there is a lot we can do to treat seizures. That is what the revolution in treating Parkinson's disease was all about. The symptoms of PD are related not just to the death of dopamine-producing cells, but to the loss of the influence of this dopamine on surviving cells of the striatum. Replacing that dopamine improves the functioning of living cells, and as a result the lives of the patients improve dramatically—even though some dopamine-producing cells continue to die off. Chorea, I realized, could be caused by some neurochemical such as dopamine acting on diseased rather than dead neurons of the striatum.

I focused upon dopamine, even though it was far from being the only transmitter that acts on the neurons of the striatum. Nor was it just that my knowledge of PD had led me to know more about dopamine than the other neurochemical messengers. What I did know about dopamine was that there was already plenty of evidence that it played a role in the production of chorea, evidence waiting to be put together.

If I knew how to treat a manifestation of abnormal nerve cell function, and knew what the mechanism of action of the drugs was that I used to treat its symptom, then by mere logic I would know something about the mechanism that produced the symptom. Neurologists think backward. Most of the medications that were used to help control chorea in Huntington's were drugs that had originally been developed to treat schizophrenia, like chlorpromazine (Thorazine) and haloperidol (Haldol). These drugs all share one property. They block the dopamine receptors of the striatum. The receptors are those areas of the membranes of neurons that are specially constructed to receive and respond to neurotransmitters. Each neurotransmitter acts upon a specific set or sets of such receptors. Dopamine acts upon the dopamine receptors on the membranes of striatal neurons, the cells particularly affected in patients in Huntington's chorea. Blocking these receptors decreases the activity of dopamine at its receptors. And that ameliorates chorea. This suggested to me that dopamine played a role in producing chorea.

But how? There was no evidence that the dopamine in these patients was being abnormally increased. So how could a normal amount of dopamine be producing an abnormal effect? Perhaps the cells' *response* to the dopamine was the key. This was a new way of looking at chorea: the possibility that chorea was not due to loss of functioning of dead cells but altered functioning of living cells.

How could I use this possibility to help patients with Huntington's chorea?

I probably couldn't.

But it occurred to me, it might make a difference to those individuals who were at risk for the disease. They knew the odds were fifty-fifty that in their thirties they would develop the disease. But many of them wanted to have children. Over the years those subjects at risk often came calling and asking for help. Could I do anything that might help them decide whether or not to have children?

There wasn't. We had no way then of detecting the gene. This was 1971, eons ago in the field of molecular genetics.

Bill Hamilton's daughter was one of those that came to me

asking for my help. Her name was Penny, and her grandmother brought her to see me, just as she had brought her son, Bill, to see me three years before. Penny was sixteen and already thinking of getting married and starting a family. There was nothing the grandmother could do to talk her out of it. Not that the grandmother blamed her; Bill's wife had fled the scene years ago as Bill became more and more irascible. Penny wanted to leave as soon as she could.

Could I help Penny? Could I tell her whether or not her kids might get the disease? She wasn't interested in her own probability, only that of her future kids.

By the time I had encountered this part of the Hamilton clan, I had come to the conclusion that it might be possible to produce chorea in those subjects at risk by increasing the amount of dopamine in the brain. At the same time, that increased amount of dopamine in the brain might not cause chorea in normal individuals. Early on it would become clear which subjects carried the gene—hence it could answer the question Penny was asking: Should she have children? Did she have the gene that would put her future children at risk?

Of course, the logic behind my thinking was not so simple as "more dopamine equals chorea in those carrying the gene." Bill Hamilton carried the gene for thirty-two years before he showed symptoms. "Thirty-one," his mother corrected me. It seems unlikely that the striatal response to dopamine suddenly shifted from completely normal to distinctly abnormal. To me this progression in the severity of the abnormal movements had to be mirrored by a parallel increase in the degree of the biochemical and physiological abnormality that triggered the movements. That abnormality in response, like the movements it produced, could not be a simple all-or-none dichotomy, but rather a continuum from normal to severely abnormal. That meant that it was conceivable that light degrees of abnormality were present within the striatum long before any chorea could ever be observed. If this physiological abnormality preceded the appearance of disease—preceded that point in time when normal levels of dopamine elicited abnormal movements—it was easy for me to believe that increasing the level of dopamine that caused such abnormally responding dopamine receptor systems might unmask this abnormality. That would be real anticipation. A way to anticipate the occurrence of the disease.

It was a hypothesis that could be tested. Increasing the amount of dopamine in the brain was easy. All you had to do was to administer levodopa. I knew how much had to be given to produce an increased level of dopamine activity in the striatum because that was what I did every day of my professional life when I gave levodopa to patients with Parkinson's disease. Once the levodopa entered the brains of these patients, it was converted to dopamine in sufficient amounts to improve their symptoms.

My hypothesis, of course, assumed that levodopa would not produce chorea in normal individuals—in this case, individuals without the gene for Huntington's chorea—whether or not that individual's family carried the disease. When given levodopa, those people at risk would then fall into two separate groups: those who would manifest chorea and those who would not. It was supposed that the former carried the gene and had developed a sufficient degree of physiological abnormality to respond to the raised level of brain dopamine, while the latter either did not carry the gene or, if they did, had not yet moved as far along the continuum of symptom severity. The appearance of the chorea during a period of increased dopamine activity within the brain would have anticipated what would naturally occur later in the course of the disorder.

I explained all of this to the Hamiltons. Penny's grandmother took charge once again. She wanted to know precisely what she would learn from all this testing. And Penny did too. After all, it was Penny's body.

I honestly admitted that I wasn't certain.

Would it answer Penny's question?

I was unwilling to give them any guarantees. I hoped that we would learn something more about Huntington's which might in the long run benefit her family.

Mrs. Hamilton persisted. I hesitated. She demanded. So did Penny, sort of.

"If you develop chorea," I told Penny, "that might mean that your chances of getting the disease would be far greater than fifty-fifty."

"But what if she doesn't? That would make it less than fifty-fifty, wouldn't it?"

"I guess it would," I grudgingly conceded.

"How much less?"

I had no idea and refused to guess. Guessing is not what scientists are supposed to do. Nor physicians. Furthermore, I am not good at it.

Penny became part of the project. She was one of the subjects at risk whom I and two colleagues, André Barbeau of the University of Montreal and George Paulson of Ohio State University, studied. We all followed the same protocol in which we gave the subjects at risk levodopa and examined them periodically to see if they developed any choreatic movements. We studied thirty subjects. One third of them developed chorea while taking levodopa. The other two thirds did not. None of the twenty-four control subjects, normal individuals who were not at risk for Huntington's chorea, developed chorea while taking levodopa. That was just what I had hoped for. Levodopa, when given to subjects at risk as we administered it in our protocol, divided this group into two separate populations. One population of ten individuals developed chorea. In all ten the chorea disappeared when the levodopa was stopped. The other twenty subjects remained free of all chorea. Penny Hamilton fell into the latter group. She never developed a single twitch. None of the control subjects, who were not at risk, developed chorea.

We were very cautious in our interpretation of the data. A positive test might not indicate that the subject would go on to develop Huntington's. All it indicated was a state of physiological responsiveness that was rare in normal controls. There is always the existence of false-positive and false-negative results.

It was quite possible that at the time of the test a subject who carried the gene for Huntington's might not have shifted far enough from normal to be detected. This would be called a false negative. We had no idea at all how often false negatives occurred.

This was the boat in which Penny Hamilton landed. She had taken the levodopa we had given to her and she had not developed any abnormal movements. Did that mean that she didn't have the gene, That she had escaped? That her children had escaped? And her grandchildren? And all of their offspring? Forever and ever?

I couldn't tell her that. I couldn't make that kind of prediction. I must have said that a dozen times. If not more. But I doubt that she ever heard it. She had looked in her mirror each and every morning of the entire two months she had been taking that medicine. Morning after morning, week after week, and she had seen nothing. No movements. That had to mean something to her and her children.

I hoped she was right.

ONCE AGAIN I lost contact with the Hamilton family. I did hear something about them, more specifically about Bill Hamilton. It was at another Sunday afternoon meeting of the Chicago chapter of the Huntington's Disease Association. At that particular meeting, someone whom I did not recognize came up to me and told me that poor Bill Hamilton had died. His house had caught fire.

"What happened?" I asked.

"It was an accident," the man replied.

"Like Woody Guthrie's mother?" She too had had Huntington's and had died when her house burned down—the result, it was said, of her having fallen asleep while smoking a cigarette.

"Yes, he was sleeping. He'd also been drinking. He was a chain-smoker. Worse now than before he took sick. He always smoked in bed. And . . ."

"And what?"

"There was gasoline in his bedroom. Spilled gasoline."

"Spilled gasoline?"

"The coroner decided it was an accident. But who knows? Maybe he just decided he'd had enough."

Suicide. It was far from rare in Huntington's. The coroner had made the generous decision. The next morning I put a brief note in Bill Hamilton's chart. Now it was between Bill Hamilton and his maker.

MY NEXT ENCOUNTER with the Hamiltons took place some time about seven years later. This time I sought them out. I was attempting to follow up the subjects at risk who had taken levodopa, which meant of course that I would contact Penny Hamilton.

Her married name was now Penny Lewis, but she was still living in the same small town, and she was willing to talk to me. She told a different story about her father's death—that he had died in an auto accident. I already knew as much as I needed to know and didn't want to contradict her. Whatever was easier for her to live with was fine with me. Her grandmother was also dead; she'd had cancer of the breast.

Penny insisted that she was normal. Just like I told her she would be. I had never said that, but that was what she had heard, what she had seen in her own mirror. Her eyes didn't lie to her. I had been through this before with others whom we had studied. The same interpretation of what had been said made up of part implication and ninety-nine parts hope.

She was too busy raising her three kids to travel to Chicago to be examined, but she would be willing to have a local neurologist see her. Someone I had trained practiced not far from her, and it was arranged that he would see her. A week later Penny was examined and pronounced free of chorea.

Once again Penny Hamilton Lewis became a statistic. By the eighth year after we had tested our last subject at risk, the two groups were different. The results were no longer all versus none—they were not that clear-cut—but they were different. Only one of the twenty in the group that had not shown any chorea while on levodopa had gone on to develop clinical evidence of Huntington's chorea. That was Penny's group. One false negative out of twenty, 5 percent with disease. During those same eight years, five of the ten who had developed chorea on levodopa had by now been diagnosed as having Huntington's. That was 50 percent. The results were not consistent enough to be a completely reliable genetic predictor, but right enough to have made our point. I was hopeful that I would never have to see Mrs. Penny Lewis or any of her offspring in any professional capacity ever again.

I turned out to be wrong about that. Penny sought me out almost fifteen years later. In terms of our knowledge of the genetics of Huntington's chorea, it was almost as if fifteen light-years had gone by, not merely three quarters of a generation. The gene for Huntington's disease had been discovered in 1983 by the Huntington's Disease Collaborative Research Group, a scientific consortium made up of fifty-eight researchers in six separate university-based research groups. The defective gene was found on the tip of the short arm of chromosome 4. In fact, the precise location on the short arm's tip is also known, but why complicate matters?

The gene is what Chicagoans would refer to as a "three-peat." Over the years, medical scientists had learned to think of an abnormal gene as a segment of DNA in which one of the building blocks is missing or coded incorrectly. Not so for Huntington's disease. All chromosomes are made up of DNA, deoxyribose nucleic acid. The complete DNA molecule is made up of two strands that wrap around each other much like a long twisted ladder. This is the double helix of DNA. The sides of the ladder are made up of sugar (deoxyribose) and a phosphate molecule, while the connecting feet are made up of a nitrogen containing base. Each strand is a linear arrangement of these three-part units (deoxyribose–phosphate–base), and each unit is called a nucleotide. In DNA there are only four different bases: adenine, thymine, cytosine, and guanine. These are fondly called by their initials, A, T, C, and G. The particular order in which these bases are arranged along the DNA strand is referred to as the DNA sequence. That is what genetics is all about, an elegantly simple system in which four different sub-parts of nucleotides are arranged in a specific sequence. Nothing could be simpler. This sequence is what determines the exact genetic instructions that will be carried to the daughter cell and on to the next generation. The two strands of DNA are held together by rather weak bonds connecting the bases on the two strands. Each pair is called a base pair. The sum total of nuclear genetic material (that total being called a genome) is calculated by the number of base pairs. The human genome is made up of about three billion base pairs. That is the sum total of our nuclear genetic material.

For a cell to divide, the DNA molecule unwinds as the weak bonds between base pairs are severed. This allows the strands to separate. Each of the separated strands then directs the synthesis of a new, complementary strand that is a duplicate of the previous companion. In this way, every time a cell divides, the genome is duplicated. I'll say it this way. Every time a cell divides, about 3 billion base pairs of nuclear genetic material are duplicated.

What then is a gene?

A gene is the basic unit of the genome. It consists of a specific sequence of nucleotide bases. These genes or units carry the information which acts through RNA to direct the building of specific proteins, which can act either as structural proteins or as enzymes. The protein is instructed through strands of messenger RNA, formed much in the way that the paired DNA molecule was reconstructed and again following strict base-pairing laws. This process is called transcription. This messenger RNA then leaves the nucleus, enters the cytoplasm of the cell, and once in place, acts as a template for the building of proteins. The genes themselves vary in length but can contain several thousand base pairs—those elegantly simple sequences. As far as we know, only about 10 percent of the genome is made up of such coding sequences.

In the construction of a protein using the right DNA sequence is critical. Proteins are themselves long chains of simple molecules called amino acids. Each specific protein, whether it acts as an enzyme or a structural protein, is made up of a particular sequence of amino acids. In humans there are over twenty different amino acids. Each one is coded for by a particular codon, which consists of a sequence of three nucleotides.

The average gene contains a sequence of one thousand codons, so that the final protein product is a predetermined sequence of about one thousand amino acids, always strung together in exactly the same order—unless something has gone wrong. The way in which things go wrong has been known for a long time. A single error in a single codon results in a protein molecule that contains an incorrect amino acid. What difference can one single amino acid make in a series of a thousand or more amino acids? The gene product in sickle-cell anemia is such an abnormal protein. The abnormality is such that the sickle-cell hemoglobin will distort the shape of the red blood cell from a disc to a sickle, alter its function, and cause it to clog up arteries, which leads to the destruction of tissue and even to death. All because one aberrant nucleotide with one mutated codon codes for one different amino acid within the hemoglobin molecule. That is sickle-cell anemia. One wrong amino acid.

That is how we have traditionally thought of genetic diseases. One wrong amino acid due to one misplaced nucleotide. That was what mutations were. But the mutation in Huntington's is different.

In 1993, ten years after the gene for Huntington's disease had been localized, the Huntington's Disease Collaborative Research Group identified the specific gene for Huntington's disease. At first they called it "important transcript 15," or IT15. Later it became known as Huntington. And here's the difference: the Huntington gene does not consist of a wrong nucleotide in a single codon. It has a repeated codon, better known as a trinucleotide repeat, or a "three-peat." Normally that gene consists of repeats of a specific codon, cytosine–adenine–guanine, the CAG codon. The research group discovered that the number of repeats was higher than normal in known patients with Huntington's disease. In other words, the mutation that produces Huntington's disease involves an unstable segment of DNA in which there is an abnormally increased number of repeats of a normal codon. Thus the mutation is not a qualitative change in the DNA sequence but a quantitative one. Three-peat mutations have also been found in a number of other genetic disorders, such as fragile X syndrome and the myotonic form of muscular dystrophy.

Why are such three-peats dominant? It is thought they arise because they produce a gain in function as opposed to a loss of one.

Each chromosome in a pair of chromosomes serves as a template for making messenger RNAs, which in turn are converted into proteins. In the more common mutation situation, the aberrant codon produces a gene product characterized by a loss of function. In other words, half of the messenger RNA coming off the two DNA templates makes a protein that cannot carry out its enzymatic function. So what? Our cells contain more copies of each enzyme than they need. Moreover, the DNA of the other chromosome in the pair is producing normal copies of that enzyme. This is why the pairing of two such mutations (one on each chromosome) is needed to produce such a gene-enzyme deficiency loss of function. The result is the classic Mendelian recessive pattern of inheritance.

But in a gain-of-function situation, the abnormal sequence produces a gene product that has some new or altered function. This protein doesn't just sit there; it does something in the cell. So even though the corresponding sequence on the other parent chromosome produces the normal product, which may well function normally, the abnormal gene also expresses its function. Hence, the mutation is dominant. If the gene always fully expresses itself, then we say there is complete penetrance of the mutation. If the product competes with the normal product, then penetrance is incomplete.

In Huntington's disease the genetic pattern of repeats is as follows:

1. Normal individuals from families with no history of the disease have an average of eighteen repeats, ranging from a low of nine or ten to a high of thirty-four to thirty-seven.
2. Patients with Huntington's disease have an average of forty-seven repeats, with a range from thirty-seven to well over one hundred.

In diseases other than Huntington's, the CAG repeat is unstable and expands further after birth, explaining perhaps the progression of that disease. But in HD the number of repeats remains stable. This suggests that in Huntington's the number of repeats is determined entirely during the production of the sperms and eggs. It is during this process that there is a tendency for further expansion of the three-peat; (that is more gain-of-function codons are added, resulting in "mutants"). And this tendency is not an equal opportunity employer. It is far more likely to occur during the production of sperm than during the production of eggs, in the father not the mother. Overall offspring tend to have more repeats than their parents did, especially when the diseased parent was the father. And the greater the number of trinucleotide repeats, the earlier the disease tends to occur.

Anticipation! With a capital A.

So anticipation was real. It was not an artifact of case finding, but part of the disease itself. Those old clinicians had it right all the time. Let's hear it for the clinicians.

So, as I was saying, Penny Hamilton Lewis again reentered my life. She brought her son with her, *déja vu*. Three generations back, remember, Penny's grandmother had brought her son to see me. Bill Hamilton—Penny's father.

Penny still looked in the mirror every morning to check for signs of chorea, but there were none. She was still "normal."

But her son, Greg, was not. His diagnosis was as obvious as

the slight shrugs of his shoulders. The occasional twitches of his fingers. The rare grimace of his face.

Anticipation! It might be more common in sperm production than in egg production, but Penny had been the subject at risk. It was her father and grandfather who had had Huntington's disease and, now painfully obvious, her son who had it. Poor Greg, eighteen years old, had anticipated his grandfather *and* apparently his mother. Using my backward neurologist's logic, it was clear to me that though Penny showed no symptoms, it might be only a question of time. From Bill Hamilton to Penny Hamilton Lewis to Greg Lewis. That was what I wrote on his chart as I constructed the family tree.

But then, why wasn't she twitching? Why hadn't her disease already started? And what about the research—would many other subjects at risk who tested negative for Huntington's disease prove us wrong? I assumed it was a question of statistics.

But perhaps I was putting the cart before the horse. Did Greg really have Huntington's? What had she noticed? As I talked to her, I watched him.

Greg thought his mother was making a mountain out of a mole hill. Nothing was wrong with him. He was anxious sometimes, but who wasn't?

Penny told a different story. Over the last year or two, Greg's behavior had changed. At first the changes had been slight, and she had tried to either deny them or ignore them or make excuses. Adolescents will be adolescents. Especially boys.

What kind of changes?

He'd always been high-strung, but he was increasingly nervous. More easily agitated. And his schoolwork had gotten worse. She stopped.

I pressed her for details.

He'd never been a great student, but he'd always worked hard and gotten reasonable grades: all B's with an occasional C.

"I got A's in Spanish," he almost shouted out.

"Not last semester," his mother contradicted him.

"I got a solid C."

"And," she added, "you flunked U.S. history and biology."

"Those teachers were creeps," Greg explained, with a shrug of his right shoulder that came too soon and was far too forceful. And was followed by a twitch on the left side of his face.

"And the movements?" I asked softly.

"What movements?" he demanded.

Penny said she had first really become aware of them in the last six months. No neurology resident could have done a better job of observation or description. As she detailed what she had seen, it became obvious that Greg had been manifesting choreatic movements for at least six months. I saw them: twitches and jerks, not quite random in location but almost; and the movements were of variable speed, amplitude, and direction and compounded by apparent attempts to transform the spontaneous unwilled jerks into almost purposeful gestures.

Greg had chorea.

Greg had undergone a change in his personality. And an intellectual decline. Not the insanity that George Huntington had described as the terminal manifestation, but the types of changes typical for the disease. Greg had a family history of Huntington's. And even though we now had the tools to prove it, Greg didn't need genetic testing. Clinical examination was enough. Greg had Huntington's.

It was her other three children who might need such testing if they wanted it. And Mrs. Lewis herself.

"Why me?" she asked.

I explained that her husband was not a subject at risk. He had no family history of the disease. She was the subject at risk, the one handing on the trinucleotide three-peats to her offspring. Or at least to one of them, and each of the others still had a fifty-fifty chance of getting the disease.

I suggested that we might continue this discussion in my office, so we left Greg by himself and walked down to my office. I sat behind my desk and she sat opposite me. This was not the first time I'd had discussions with female members of her family in this office. They were not meetings that I ever looked upon with warm anticipation. Back to that word again. That was why I wanted her to be tested. Not for her benefit, but for mine. The scientist in me wanted to know how many three-peats she had. The clinician in me now assumed that the number of CAG codons in her gene would not tell me anything I didn't already know. She had at least thirty-seven, at least one too damn many. After that it was all superfluous. The number would only predict when her disease would start on the average. Averages meant nothing. Still I wanted to know. I had never encountered this situation before. Anticipation I had observed; after all, I am and always have been far more a clinician than a scientist. So I had seen numerous patients in whom the age of onset was earlier than it had been in the affected parent; but I had yet to see the opposite, a patient in whom the onset of symptomatic disease was later than the onset in the parent.

Penny Hamilton Lewis had no clinical evidence of chorea. I had been watching her for over half an hour. The difference in the number of three-peats between her and her son would be most interesting.

But she was not interested in my curiosity. So back to science, the clinical facts. "Your grandfather's disease began while he was in his late thirties," I began.

She was well aware of that.

Also that her father's disease had begun in his early thirties and her son's in his teens. Yet she still remained symptom-free.

She knew that; she had told me that when she had walked into my office. She was not the patient, she reminded me. Greg was.

I asked again if she would let me test her. Once again I was rejected. I explained the scientific issues, to deaf ears. I begged. I cajoled.

I realized I needed to try a different tactic. I brought up her other children. They were at risk. They should be tested.

"Why?"

That one-word question covered a profound issue. Subjects at risk often told me that they wanted to know whether they were carrying the gene in order to decide whether they should have children. On the face of it that seems to be a fair request.

But we have yet to know how often they act on their knowledge. Does knowledge really alter behavior? And if so, how
often? So Penny was saying there was no reason for her kids to
know. Maybe they wouldn't want to know. Maybe they would-
n't have kids of their own, or maybe they would. Maybe knowing the future might not change their behavior. Maybe not. But
it sure as hell could change their lives.

This issue is difficult to resolve. Nancy Wexler, a subject at
risk, has called this dilemma the Tiresias complex. She views
this in part as a subject at risk and in part as a trained psychologist. Why make a diagnosis of a disease that cannot be treated? This information leads to no solution since there is nothing
that can be done. It's part of the scientific arrogance of the late
twentieth century that diagnosis must of necessity lead to treatment and that treatment is the justification for diagnosis. Enter
Tiresias, the blind soothsayer in Greek mythology. He appears
in both Sophocles' *Oedipus Rex* and Homer's *Odyssey* as well
as several other Greek tragedies. In *Oedipus Rex,* Tiresias
knows the truth but he also knows what disasters will follow
revelation. So he tells Oedipus, "It is but sorrow to be wise,
when wisdom profits not."

Wexler has restated the dilemma with these questions: Do
you want to know how and when you are going to die, especially if you have no power to change the outcome? Should
such knowledge be made freely available? How does a person
choose whether or not to learn this momentous information?
How does one cope with the answer?

But let's not forget what happened in *Oedipus Rex*. The King
realizes that he has killed his father and married his mother. His
wife/mother then kills herself, he blinds himself, and the family
tragedy then descends upon the next generation. Had Tiresias
done the right thing? Why had Oedipus started his quest for
knowledge? Because Thebes, his Thebes, was being visited by a
terrible plague. That plague had been visited upon Thebes
because a sin offending the gods had to be corrected. The sin of
Oedipus. When the blinded Oedipus departs for Colonnus, the
plague is lifted. Did his knowledge prove to be of no value? Did

not more people profit from his knowledge than suffer from it?

All of this ran through my mind, but none of it did any good. Penny was as stubborn as her grandmother had ever been—or was it her father? "But think of your grandchildren," I blurted out.

"Don't worry about my grandchildren," she spat back at me. "Why not?"

"None of them will ever get Huntington's," she shouted.

"What about your other children?"

"Nor will they."

Her faith knew no bounds.

"How can you be so damn certain? Your son has the disease. So do you."

"Go to hell," she said.

"I . . ." I stopped. I had pushed too hard, losing sight, in my quest for science, of what was her real-life tragedy.

I paused. "Look, Mrs. Lewis, I am truly sorry that I pushed you so hard, but I was really thinking of the next generation. Facts are facts. Your father had Huntington's. Your son has Huntington's. You have Huntington's."

"I do not!" Firmly this time, almost without emotion.

"How can you be so sure?" also asked with as little emotion as possible.

"Because he is my father's son."

"Your father's . . ."

"The son of a bitch raped me." Remnant of George Huntington's allusion to the terminal insanity of Huntington's perhaps.

"He died," I recalled.

"When I was six months pregnant, he died in an auto accident."

I sure as hell was not going to contradict her. How much had anyone else known? Who had spilled the gasoline. Had it been a suicide, an accident, or something far more sinister? Had he been murdered? Maybe that gasoline had not just happened to spill. It may have been started by a cigarette, or it may have not. Was he just sleeping off another drunken binge? Had he just

finished sleeping with his daughter? Who had killed him? The
women in the family were the strong ones, the willful ones, the
ones who took decisive actions.

These questions were not mine to pursue. I was her doctor—
the past was done. I had the answer I needed. Penny was right.
Greg's disease didn't mean that she had the gene. Greg's aberrant genetic makeup came from his grandfather. Not his mother.

I might have guessed this possibility if I, like most of us, did-
n't tend to block out what's unpleasant.

There is an old tried-and-true clinical observation, one that,
unlike anticipation, has stood the test of time. In the vast
majority of Huntington's patients in whom the disease begins
before the age of twenty, the patient has inherited the gene from
the father. Three-peats galore. And Greg was under twenty.
Penny was absolutely right. She didn't have Huntington's. She
was not a false negative, although in some ways her life might
have been easier if she had been.

I started to say something but stopped myself.

Penny was also wrong. She was not out of the woods yet, nor
were her other children. As the daughter of a man who'd had
Huntington's, she was still a subject at risk. She still had a fifty-
fifty chance of carrying the gene and ultimately developing the
disease. And her other children were not out of the woods
either. I was back to the dilemma of Tiresias. But not quite.

"Who else knows about . . . ?"

"My grandmother knew."

That went without saying. Many mothers of abused daughters don't know or don't want to know—but Mrs. Hamilton
was not that kind of woman.

"Your husband?"

She shook her head. That was why she couldn't be tested. If
she was tested and they found out that she was negative, then
. . .

"But they wouldn't have to know."

"What do you mean?"

"If your test is negative, only you and I will ever know."

"What good will that do?"

"Then we can test your children."

"But they'll all be negative . . ."

"That's the entire point. Right now, the sword of Damocles is over their heads. But they are probably not at risk—so they should know."

And that was what we did. Penny had twenty repeats on one chromosome and fourteen on the other. She did not carry the gene. When her other three children were tested, all turned out to be normal. Three out of three with no three-peats. They no longer had to worry. Their own children would not be threatened. There was no Huntington's disease in their future.

11

THE HERMIT OF
THIEF RIVER FALLS

·

On First Meeting an Eponym

WHENEVER A FAMOUS visiting professor comes to visit any hospital's department of neurology, one of his or her obligations is to give a lecture. These talks are usually given during grand rounds, a format in which a patient is presented and then the VP (visiting professor) uses that patient as the starting place for his or her talk. This arrangement leaves the host with numerous obligations, the most important of which is to choose the right patient to present, one that would allow the VP to wax eloquently on a subject he or she knows much about. Most lecturers are sponsored by drug corporations and paid handsome honorariums. Usually, the VP uses the same speech, taking it around from one grand round to another. That's how the system worked in 1964—the year this tale took place—and still works now, with one exception; now the honorariums are larger.

The VP I write about here had already become an eponym, but the disease named after him was quite rare. There weren't more than a couple dozen cases in the entire country, so unless he brought the patient with him, most neurology departments would not have been able to present the right patient to him. That is one of the tragedies of twentieth-century neurology. It is still possible to become an eponym but all of the good diseases are already taken. Only the rare diseases are left, and it was one of the rarest of the rare that took on the name of our VP, Professor Sigvald Refsum of the University of Oslo. He was the first eponym to come out of Norway since Armauer Hansen gave his name to the bacillus that causes leprosy—Hansen's bacillus—and later to the disease itself, Hansen's disease. But the latter was done not so much to honor Hansen as to find a name for leprosy that was free of the stigma attached to the traditional name. Much like attaching J. L. H. Down's name to mongolism. Lepers are no longer called lepers; they have become "patients with Hansen's disease."

There was no stigma at all to having Refsum's disease. The Hermit of Thief River Falls did not become a hermit because he had Refsum's disease. In fact, as I learned from Professor Refsum that day, it was just the other way around.

Thief River Falls is a small town in rural Minnesota, and Knut Jacobsen had lived there all of his life. He was a farmer just as his father and both of his grandfathers had been. The families had immigrated from Norway in the late 1880s, and settled on land near Thief River Falls. They were hardworking farmers who still felt attached to the Old Country. Knut was named after Knut Hamsun, pen name for the Norwegian writer Knut Pedersen, who won the Nobel Prize for Literature in 1920, the second Norwegian writer to be so honored.

The family had always been independent and had stayed fairly much to themselves. Then in the late 1930s, when Knut was in his middle teens, his parents and older brother had died in a terrible fire and he was left on his own. For a couple of years he lived on the farm, or what was left of it, until creditors started claiming various parts of the land. By the time the United States entered World War II, Knut was living in a small log cabin on a small lake deep in the backcountry and supporting himself by catching fish and hunting.

He received a draft notice, which he ignored until the sheriff dragged him into town. The draft board decided, based on psychiatric grounds, that he wasn't fit for service. Knut returned to the log cabin and ever-increasing isolation.

World War II meant a lot of things to Americans. One of them was rationing with blue points and red points, the latter for meat. Knut Jacobsen had no need for such things. Venison wasn't covered by the rationing system. Nor were rabbits. Nor fish. He could eat all the protein he wanted. So could any other hunter or fisherman. But most of the people who lived in Minnesota weren't fishermen or hunters. They were farmers, shopkeepers, teachers, barbers, truck drivers, and factory workers. They all had fewer red points than they needed. All of a sudden there was a market for anything Knut could catch or kill, and he soon had a reputation for supplying the best fresh game in all of Minnesota. It was always dressed perfectly, not an ounce of fat on it. Once a week Knut made his way into Thief River Falls and sold his game, already dressed, and his fish, always for cash, but he never bought anything. Knut lived off the land and was totally self-sufficient. As the war went on, he took in more and more money. But good things never last. D-Day was followed by V-E Day and then V-J Day, and then within a year or so rationing ended. No more red points meant that no one needed Knut's fish and game. And then the game wardens returned, and they actually took the deer season seriously.

So Knut stopped fishing and only hunted during the season. He would come into town once or twice a year, but otherwise he kept to himself out in the woods. By 1960, the townsfolk of Thief River Falls began to notice that something had gone wrong with old Knut. Though he wasn't yet thirty, he looked forty. His skin was thick and hard and looked like the scales of some dying fish, strange even for a man who had lived out-of-doors all of his life. And Knut's walking changed. He had started to drag his feet as if he could hardly pick them up. Then he began to stagger, as if he were drunk.

Could that be how he spent all that money he'd made during the war? Buying bootleg alcohol? Alcohol had required no points of any kind. No one in Thief River Falls had ever sold him any alcohol, and no one had ever seen him drunk, but who could know for sure? What else was there to do on those long, cold winter nights in that lonely cabin in the back woods? Perhaps he made his own liquor. Or bought moonshine from some other hermit.

Then in early spring of 1964, Knut came to town and he looked worse than ever. He could hardly lift his feet, and his balance was so poor that he seemed to weave from side to side as if he were looking for a wall to help hold him up. It was then that the store owner, a man named Henrik Nordgren, suggested to Knut that he see a doctor, and Knut agreed. If Knut was agreeing to see a doctor, thought Nordgren, then the Hermit must believe that he was next to death's door.

That was how he became my patient. Henrik Nordgren brought him to the emergency room at University of Minnesota Hospital. It wasn't that this was an emergency—it had been going on for many years—but Nordgren figured if he waited to schedule an appointment, Knut might change his mind.

The doctor in the emergency room took one brief look at Knut Jacobsen and called for help. He paged the dermatology resident on call and the neurology resident on call. I got there first. Knut was lying on his back, staring at the ceiling when I walked up to the bedside and introduced myself.

"Never met no new-rologist before," he said. "Of course, I don't meet too many people these days." He paused and then asked, "What is a new-rologist?"

"Neurologist," I corrected. "We study the nervous system. We try to help patients who have problems because their nervous system might not be working right."

"What sort of problems?"

I looked down at the brief notes the emergency room physician had written down on his record. "Like trouble with walking."

"What sort of trouble? Rheumatism?"

"No. Dragging of the feet. Loss of balance."

He said nothing. I looked down at the sheet again.

"People with a drunken gait," I offered.

"I never drank a drop in my life."

"No one ever said that you did. But you can walk like you're drunk without ever having had a drink in your life."

"How's that?"

I was adjusting to his terse questioning. "That's what neurology is all about." I paused and when he didn't say anything, I went on. I explained that alcohol caused imbalance by affecting the function of a part of the brain called the cerebellum.

He asked me to repeat the word, which I did. He said it out loud.

The cerebellum, I explained, controlled balance. Alcohol altered that control, but so did other things including any number of neurological diseases, and my job would be to find out what was causing his cerebellum not to be working normally and, if possible, to fix it.

"You don't do surgery, do you? Brain surgery?"

"No. No surgery. That's done by neurosurgeons. I'm a neurologist."

"A neurologist is what I want then," he said pronouncing the word perfectly. "And I want you to be my neurologist."

I tried to explain that I was just a first-year resident. That I wasn't a real neurologist yet, and that he would be assigned to one of the attending neurologists who had far more experience than I did.

Experience didn't matter that much to Knut.

And I wasn't even on the admitting service. I was merely doing night call; during the days I was in the laboratory reading electroencephalograms.

That was another word he repeated.

"Brain-wave tests," I explained, although I'm not sure the explanation meant anything to him.

"Sounds pretty boring to me."

It was very boring, but I was only a first-year resident and had to do what I was assigned.

"I'm sure you can find time to see me in between brain-wave tests."

I couldn't do that.

"Then I guess I'm going back home."

"Then I guess I'll find time to see you."

I hadn't even taken his history or examined him, yet already he had drafted me into service as his doctor, a role I continued to play long after I left Minnesota far behind me. I pieced together his history from talking to both Knut Jacobsen and Henrik Nordgren. It was Mr. Nordgren who told me that Knut was known as "the Hermit" on account of his being a loner who lived all by himself in a log cabin deep in the backwoods. Mr. Nordgren also explained that Knut made his living as a hunter and fisherman and was almost a legend in his area.

As I took notes of our conversation, I began referring to Knut as the "Hermit of Thief River Falls."

Mr. Nordgren continued to give me more details of the progression of Knut Jacobsen's disability, since Knut was unable to remember when had it started.

Mr. Nordgren, however, thought he had noticed that something was wrong in the middle 1950s, when Knut was in his late twenties.

What had he noticed first?

Some dragging of his feet.

Slowly a coherent story emerged. Knut had been dragging his feet for about a decade. Both feet dragged just about the same, and the dragging had slowly but inexorably gotten worse over the years.

The dragging of his feet suggested weakness that originated distally, removed from the actual location of the disease, and then progressed. Since the dragging was equally evident in both legs, it seemed likely that Knut was suffering from a peripheral neuropathy, a disorder of the nerves that innervate the muscles of the legs. This also suggested he had a dying-back neuropathy, which meant that the nerve cell bodies were diseased, not simply the nerve endings. The cell bodies, which are located within the spinal cord, control all of the metabolism of the nerves. If the cell bodies don't function normally, the cells begin to die, and they die "backward," starting at the far end. A dying-back neuropathy, then, begins in the feet and works its way up the legs.

Such distal neuropathies can be caused by poisons or toxins (such as lead, arsenic, or thallium), by metabolic problems, (such as diabetes mellitus), by hereditary diseases, or by nutritional or vitamin deficiencies, which often occur in alcoholics.

Anyone with a dying-back neuropathy of both legs for ten years should also be showing some neuropathy of his hands; they too should be weak. Were Knut's hands weak?

He thought so.

I looked at his hands. The muscles of his hands were atrophied, which was also consistent with this pattern of neuropathy.

So the Hermit had a diffuse dying-back neuropathy of all four extremities.

Over the last three or four years he had been walking as if he were drunk.

"I never drink a drop," he protested.

Was he protesting far too much?

Most alcoholic neuropathies are more sensory than motor, that is, they involve sensations rather than actual movement problems. More numbness and tingling than dragging of the feet. Still, I had been taught that by far the most common cause of peripheral neuropathy complicated by ataxia, or imbalance, was alcoholism, followed closely by alcoholism as the second most common cause with alcoholism also coming in as a close third. The Hermit was probably a closet drunk living in a log cabin. My patient indeed! I already had more alcoholics on my roster than I wanted. Their main virtue as patients was that they rarely showed up for their appointments giving me more time to spend in the library.

Knut was presenting another symptom. Over the last year he had developed trouble with his vision in both eyes.

Was the problem the same in both eyes?

He thought so.

What kind of problem?

Everything was blurred and indistinct.

Had it started all of a sudden? Bad booze, booze made with wood alcohol, or methanol, can cause blindness by injuring the optic nerves. That usually begins suddenly.

No, it had been gradual.

Probably it was not methanol poisoning then. I looked into his eyes, examining the eyegrounds (the backs of the eyes). I did not see a white, shrunken, optic nerve, the telltale sign of methanol poisoning. Instead, his entire retina was abnormal, covered with black spots of various sizes and shapes as if someone had sprayed some sort of black pigment over it and caused it to be inflamed. I had never seen such a situation before.

I then performed the rest of my neurological examination. I watched him walk. He had what is called a steppage gait. His feet, really the muscles of his lower legs that flexed his ankle, were so weak that as he stepped forward he had to lift his knees as if he were going up a step in order to get his toes off the ground. This I understood. The muscles of his ankle were very weak but those of his hip were strong enough to compensate, which confirmed that he had distal weakness—just what I had predicted.

His steppage gait was the same on both sides: perfectly symmetrical, bilateral distal weakness that went along with the bilateral atrophy of both hands. I had him take off his shoes, (he wore no socks), and lift his trousers. The muscles of his lower legs and feet were atrophic, like the hands.

Knut also had loss of sensation in his extremities. He couldn't feel a light touch on either foot, but could at the knee. The same for a prick with a pin and the vibration of a tuning fork. Then with his eyes closed, I changed the position of one of his toes and asked him if I had moved it up or down. He was unaware that I had even moved it. He could tell if I moved his ankle and even got the direction correct, most of the time. Significant sensory loss that was distal and symmetrical was never part of muscle disease or amyotrophic lateral sclerosis (Lou Gehrig's disease). Nor was whatever I had seen in his eyes.

Knut's nutritional status looked good. Other than his atrophic musculature, he was rather robust. He probably had a reasonable diet. I asked him what he ate. Fruits and vegetables galore. So much for a vitamin deficiency.

That left Knut Jacobsen with a peripheral neuropathy and funny-looking eyegrounds. And he was ataxic, walking with imbalance.

Peripheral neuropathy, ataxia, and funny-looking eyegrounds.

It was at that juncture that the dermatology resident walked into the room.

"What do we have?" he asked.

I introduced him to Knut Jacobsen. I didn't know the derm resident, and I was certain he was never going to introduce himself. Residents being the way they are, he stood several feet away and looked at the patient. He followed the basic rule of dermatological examination: If you don't know what kind of rash you are looking at, there is no reason to touch the patient; rather, observe, observe, observe.

"What do we have?" he reiterated.

"I do not know what we have," I replied. "But Mr. Jacobsen has had this skin problem for five or more years. His skin had been getting increasingly scaly." That was an understatement, his skin looked like that of a fish.

Knut held out his arms and showed his back and chest, so the dermatologist could examine him.

"He's got fishlike skin," the dermatologist said.

"I know that much. What's he got?"

"Ichthyosis." But that simply meant fishlike skin.

"He has ichthyosis," he reiterated. "And there is no treatment, and it certainly is not an emergency." With that he walked off.

Peripheral neuropathy, ataxia, funny-looking eyegrounds—and ichthyosis. Well, if the derm resident could call Knut Jacobsen's fishlike skin "ichthyosis," then I would call his inflamed retina "retinitis." "Retinitis pigmentosa," retinitis with splotches of pigment. Big deal. I knew Latin, so what?

Then something clicked. Knut Jacobsen had Refsum's disease, named after Dr. Sigvald Refsum.

Dr. Refsum had originally given the disease the perfect Latin descriptive name: heredopathia atactica polyneuritiformis, which translates to a hereditary disease *(heredopathia)* characterized by ataxia *(atactica)* and a neuropathy of several peripheral nerves *(polyneuritiformis)*. And the patients usually had retinitis pigmentosa and often ichthyosis. It was a high school Latin teacher's dream come true. And not a bad diagnosis for a first-year neurology resident.

I admitted him to the hospital. We didn't do very much for him. All of the senior neurologists came by, nodded sagely, and agreed with the diagnosis. Of course, none of them had ever seen a patient with Refsum's disease before. They all called it that. No one ever said "heredopathia atactica polyneuritiformis." Refsum's name was easier to pronounce and even easier to remember, and besides, Refsum was coming to give grand rounds in June. I could present my case to him then. That meant I had three months to get prepared.

But what could we do for my patient now?

None of the staff neurologists had any idea what to do, or even if there was anything we could do. They all told me to let Professor Refsum tell us. He'd discovered the disease. He'd know what to do with my case.

But Knut Jacobsen wasn't a case. He was a patient, a person, a man who's nervous system was slowly deteriorating. The key word for all of my mentors was "slowly." That meant that, whoever the patient was, he could easily wait three months.

What choice did he have?

So we waited. Mr. Jacobsen waited back in his log cabin in the backwoods beyond Thief River Falls, fishing by himself and seeing no one: roles that befit the Hermit of Thief River Falls. He no longer hunted. His vision was not sharp enough. He hadn't shot anything in years. He couldn't even put out traps anymore because his hands were too weak to open them. Mostly he eked out an existence by gathering whatever he could off the shrubs and other plants in the woods.

I, too, waited, but mostly in the library. In those three months I read every article I could find on Refsum's disease. I started with Refsum's original article, published in 1946. Refsum had written about four patients he had studied for the last three years of the Nazi occupation of Norway. The patients had come from two different families, two from each family. That was how he knew it was a hereditary disease. The two families lived at opposite ends of Norway, over five hundred miles apart. As far as anyone knew, the families were not related. None of the parents of the four patients had the same disease. Nor had their parents before them. That meant that the most likely mechanism of inheritance was as a rare recessive gene. When a gene is recessive, an individual with one allele, or copy of the gene, is functionally normal but can pass that allele on to his or her offspring. If the other parent is not a carrier (that is, has no alleles for the gene), then that passage doesn't matter: Half the children of such a union will have one copy of the gene, the other half no copy of the gene. But *all* of the children will be normal, because individuals born with only one allele for Refsum's disease remain normal throughout their lives. If both parents are carriers, then the chances of any one child inheriting both recessive alleles, and thus having the disease, are only 1 out of 4.

So rare remains rare.

But marriage is not necessarily a random event, especially not in the backwoods of northern Norway, where the same families have been intermarrying since the days of the Vikings. It is not the background incidence of the disease in the entire population that predicts the overall incidence of the disease, but the incidence within specific subgroups. The gene for Tay-Sachs disease is so rare in the general population that it should be a very rare disease, but it isn't. The gene is all but limited to Jews who trace their origin to Eastern Europe, who tend to marry other Jews who trace their origin to Eastern Europe. The same applies to sickle-cell anemia in the African American population. If a black person is more likely to marry another black person, and both possess the recessive sickle-cell trait, then there is an increased chance that the offspring of such a marriage could develop sickle-cell anemia.

Refsum's four patients all had the same constellation of clinical problems. That was what had allowed him to isolate them from all other patients, to determine that they had the same as-yet-undescribed disease. A familial disorder awaiting discovery, a syndrome awaiting an eponym. Reading their stories was almost like rereading my notes on Knut Jacobsen: Progressive difficulty with movement, with legs becoming involved before arms. A steppage gait with distal weakness and atrophy of the muscles, more in the legs than in the arms. Sensory loss—they were oblivious to touch, pinprick, position sense, vibratory sense, temperature. Each of the four had a peripheral neuropathy. Each of them also walked as if drunk. Each had a history of progressive loss of vision. When Sigvald Refsum looked in their eyes, he saw retinitis pigmentosa in each of them. And finally, they all had ichthyosis.

Even after Knut Jacobsen's ancestors emigrated to Minnesota, they were more likely to meet and marry other Norwegian immigrants, so it makes sense that his parents were carriers of the recessive genes for Refsum's disease.

The study of the four cases was enough to identify a new disease, but Refsum hadn't stopped there. One of the four died and had then been studied. The nerves were filled by an accumulation of fat. To Refsum that meant that heredopathia atactica polyneuritiformis was an inborn error of fat metabolism, implying that the recessive gene caused an error in cellular metabolism. In Refsum's time, no one was quite certain what the error was; that knowledge would come later. In the 1940s each gene was thought of as controlling one protein, most of which acted as enzymes. Ergo, one gene—one enzyme. Each abnormal gene resulted in an enzyme that could not perform its normal function. The loss of that metabolic function resulted in a block along a single metabolic pathway, hence an inborn error of metabolism. In Tay-Sachs disease, the inborn error was in one pathway of fat metabolism. As a result, the neurons of the brain ballooned and choked to death from the fat that had accumulated because one of the enzymes needed to rid the body of that fat was not working.

But what fat? And what metabolic pathway? Those were questions that Refsum was smart enough to raise, but for which he had no answers in his initial publication.

After Refsum's report, a slew of papers had followed. The disease was not limited to individuals of Norwegian descent. Not every patient had ichthyosis. But they all had peripheral neuropathy and ataxia and some degree of retinitis pigmentosa. The neurologists all agreed that Refsum had described a new clinical entity and that he had given it the perfect descriptive name, but the name was too cumbersome and Latin was no longer the universal language of science; English was. Refsum's original paper had been in English. So they called it Refsum's disease, and Sigvald Refsum became an eponym within his own lifetime, something James Parkinson had never achieved.

Was that why Refsum had given the disease an all-but-unpronounceable, convoluted albeit descriptively correct Latin title? I took note of this: if I was ever to describe a new disease, I should give it a terribly complex, yet accurate, Latin name. That would guarantee my becoming an instant eponym. Fame that would last far longer than fifteen minutes.

Now that I had completed my education regarding Refsum's disease, I could hardly wait for Dr. Refsum's arrival.

Grand rounds were on June 5, 1964. Knut Jacobsen was readmitted to the hospital two nights before so I could reexamine him and be ready for my presentation. When I walked into his room to greet him, I hardly recognized his voice. His speech was slurred and soft, most likely a result of weakening muscles associated with speech, and it was obvious he couldn't see me. He told me all he could see was an outline of someone framed in the light of the doorway. Knut could hardly stand without falling down, much less walk on his own. I was shocked at the drastic progression of his symptoms. He'd had Refsum's disease all his life. Why was he deteriorating so quickly? I remembered that one of Refsum's first four patients had progressed rapidly and died. Was that what was in store for Knut?

I hoped Sigvald Refsum could answer these questions.

The format of such grand rounds is written in stone. The process is designed to allow the visiting professor to shine. It's like playing an entire basketball game just to allow Michael Jordan to take the last shot with only one second left on the clock and the Bulls down by one point. I would give the history, then the patient would come in and Professor Refsum would examine him. He would get to demonstrate what he had found that had led him to make the diagnosis in his original patients and then he would tell us about his disease. I was not as excited as I had been three months earlier. I no longer believed that he would be able to help Knut Jacobsen.

Grand rounds was held in an auditorium. Professor Sigvald Refsum sat in the front row next to my mentor, Abe Baker. Refsum was a slight man. It was hard to tell his age. He was probably about sixty, though there was a possibility he was one of those people who had looked about sixty for ten years and would continue to look that way for another fifteen years. I took a deep breath and presented the history. It took me less than five minutes. I outlined all the salient points, avoiding such words as "retinitis pigmentosa" and "ichthyosis." I had slides of his skin and eyegrounds and would project them when I was told to do so by Professor Eponym. I turned to get Mr. Jacobsen, who was waiting just outside the auditorium, when Professor Refsum interrupted me.

"I understand that this patient has a nickname?"

"He does. He's called the Hermit. The Hermit of Thief River Falls."

The VP nodded and waved me on. I brought Mr. Jacobsen in. He was in a wheelchair. I had told him what to expect. All Knut wanted to know was whether this doctor could help him.

Professor Refsum introduced himself. He was the most unassuming VP I had ever watched on grand rounds. In the most gentle way imaginable, he started asking Knut Jacobsen questions—not about his symptoms, not about his family tree or his family history of possible neurological symptoms. None of the questions that I expected. He asked about his diet—in excruciating detail.

Knut lived on vegetables.

What vegetables?

Leafy vegetables. A major source of phytanic acid, I thought to myself.

And what else?

Nuts. Another rich source of phytanic acid.

I knew where Refsum was going. He was figuring out what in Knut's diet had caused phytanic acid to accumulate more rapidly over the last three months.

Did he eat any meat?

None at all.

But he was a hunter and a fisherman.

He hated fish. He sold them off.

And the meat?

He could not hunt anymore, but he had stored meat in enormous quantities. His freezer was a concession to what we would call civilized society. But he dressed it and sold it.

And what did he do with the waste? With the fat that he cut off?

That he ate. But that wasn't meat.

Dr. Refsum was masterful. In fifteen minutes he had outlined the entire history of the Hermit's peculiar eating habits, a dietary history none of us had ever bothered to elicit. A diet choked full of phytanic acid.

There were two distinct possibilities for why the phytanic acid, a fatty acid, had accumulated. Either the body was producing too much of it, or it was unable to metabolize, or break down, normal amounts, which then accumulated. The odds were on the latter. That was what the inborn error of metabolism did. One enzyme in some obscure metabolic pathway was abnormal. In this case it was phytanic acid, which then collected in the nerve cells and the skin and the retina.

Could that be proven? And where did phytanic acid come from? How was it metabolized, and why was it just phytanic acid that accumulated in Refsum's disease?

The human body doesn't make phytanic acid from scratch. It comes from food. Most of it derives from phytol, which is related to chlorophyll and is found in leaves, leafy vegetables, and fruits. Only roots, those parts of plants that stay completely out of the sunlight and have no need for chlorophyll, such as potatoes, are devoid of phytol. Phytanic acid itself also gets stored in vegetables and in nuts especially. It gets stored in animal fats too, but no one eats enough animal fat to accumulate enough phytanic acid. Most fatty acids are metabolized by enzymes that start at the "head" of the molecule. Not so phytanic acid. It's metabolism starts at the back end, which is called omega oxidation.

I returned my attention to the grand rounds presentation. Next, I assumed, Dr. Refsum would examine Knut and confirm the diagnosis.

None of that ever happened.

"Do you know what you have? What your disease is called?" he asked Mr. Jacobsen.

Knut nodded. "I have Ref . . ."

"You have heredopathia atactica polyneuritiformis," he said. And that set the tone for the rest of his discussion. Never once did he say "Refsum's disease." Never once did he put his name to the condition. Always the original descriptive name. "And," he added, "Dr. Klawans made the diagnosis, so he should be the one to examine you." With that he sat back down and I had to demonstrate the examination.

I quickly stood and approached my patient. I showed then the muscle wasting and sensory loss of his peripheral neuropathy, and by walking a few steps, his ataxia was evidenced. And then I showed the slides of his skin changes and eyegrounds. There were no other questions, so I started to leave, wheeling Knut along with me. Once again, Professor Refsum stopped me. "Mr. Jacobsen should stay so that he can hear exactly what can be done to help him."

To help him?

Refsum began his formal discussion, but this was not a canned talk that he had used across the United States from Boston to New York to Philadelphia to Baltimore to Chicago to Minneapolis. He started by congratulating me on my clinical acumen and for allowing him to participate in the care of my patient, the most important patient he had seen with heredopathia atactica polyneuritiformis since the first four who had made it possible for the disease to be recognized.

Knut Jacobsen had proven what Dr. Refsum had always suspected. The doctor went on to explain that he had not been in the right place at the right time when he was studying his four patients. Nazi-occupied Norway during World War II was in a period of harsh deprivation. Often the population was bordering on starvation. For many it was time "when the moon was down." I understood the allusion to John Steinbeck's short novel about the Norwegian resistance, *The Moon Is Down*. Professor Refsum's four patients had all had to change their diet because of the restriction placed upon them by the Nazi occupation, by the Norwegian puppet government led by Vidkun Quisling, another Norwegian who had become more than a mere eponym; he was a full-fledged adjective. Quisling. Refsum all but spat that name out at us, as if even saying the name was a painful experience.

During the war, all of Norway's fish had been shipped to Germany, and like Mr. Jacobsen the population had been forced to become a country of vegetarians. That was why Knut Jacobsen was important. He too had changed his diet. Lots of chlorophyll. Lots of phytol. For the two particular families many miles away and years before, that meant the rapid progression of their disease as they had become overwhelmed by their dietary excess of phytol and phytanic acid. Just like Knut Jacobsen. History confirmed that a change in diet could bring on the rapid onslaught of this usually indolent, hereditary disease.

Well, if the wrong diet could set off heredopathia atactica polyneuritiformis, then the right diet might just reverse the entire process.

I looked at Knut. He was smiling. He had understood—maybe not every word, but enough. He could be helped by a change in his diet. No more leafy vegetables. No more nuts. Only peeled fruits. No more animal fat. Lots of tubers. Roots. Bring on the potatoes.

It had been a brilliant discussion. Scientific and personal.

As I wheeled Knut Jacobsen back to his room, he was most excited at the possibility that he might improve. But, I reminded him, it would take a long time for his body to get rid of all the excess phytanic acid.

Time was one thing he had a lot of. Then he said, "They shot the bastard."

"Who?"

"Quisling. The Norwegians shot him. Right after the war. They tried him as a collaborator and shot him. He deserved it. Can you imagine collaborating with those Nazi SOBs?"

The next day was June 6, 1964, the twentieth anniversary of D-Day. I bought a box of cigars and passed them out. No one could understand why I was celebrating the anniversary of D-Day. I was giving a cigar to Knut when Professor Refsum walked into his room and I offered one to him.

"I never smoke cigars," he said. "But I will smoke this one. I will smoke it with pleasure."

Then in great detail he outlined the diet to Mr. Jacobsen and what he could expect. According to Professor Refsum, he would stop getting worse immediately, but improvements would take longer.

How much longer?

That depended upon how much fat was inside his nerves. Perhaps a year. Perhaps longer.

"You ought to know. It's your disease," Knut said as he lit his cigar.

As we walked out of the room, I told Refsum what Knut had said about Quisling. He handed me a book on the neurological examination. Had my exam been so bad that he thought that I needed to read another book on the subject? No, my examination had been perfect.

It was a book that he now edited. It had originally been written by Monrad Crohn, who had held the Chair in Neurology at the University of Oslo before him. When Refsum inherited the Chair, he also became editor of the book. It came with the job. He gave one to each resident who presented a patient to him. Besides, he insisted, it was a very important book for me to own.

Why?

Because it contained a normal pneumoencephalogram.

That statement puzzled me. Pneumoencephalograms are a method of obtaining an image of the brain by injecting air into the spaces around the brain and then taking X-rays. It was crude and painful, but it was all we had. Every textbook had a normal pneumoencephalogram. What made this one so important?

"It's Quisling's. We had to make sure that his aberrant behavior was not due to any neurological disease. We did a pneumoencephalogram of his brain. It was normal."

"Then he was shot?"

"Yes."

"Before or after his headache went away?" Headaches were common after pneumos. The air irritated the linings of the brain. The headaches often were quite severe and sometimes lasted for weeks and weeks.

"I never asked."

Knut went back to Thief River Falls the next day. He stabilized, and about a year later he began to improve. He never got back to normal, but no one had ever expected that. He improved to the point where he could do pretty much whatever he wanted to do. Further improvement would have required him to give up on nuts completely. That was one pleasure he was not going to give up.

SIGVALD REFSUM AND I met again some six years later when I became a member of the World Federation of Neurology's Research Group on Huntington's Chorea, of which he was already a member. He asked me about Mr. Jacobsen. I told him the diet had worked but that he refused to follow it as strictly as he should.

Sigvald nodded. The diet was difficult to follow.

Whenever we met, he asked about my patient. Then in 1980 Knut died of a heart attack. Fourteen years after that Sigvald Refsum died. I still have the book he gave to me. We don't do pneumoencephalograms anymore. I haven't ordered one in well over two decades. No matter. I still tell everyone that the second edition of Monrad Crohn's book on the neurological examination is a classic because it contains a normal pneumoencephalogram. The normal pneumo of Quisling.

And I never learned whether they waited for Quisling's headaches to go away before they shot him.

1 2

MAD COWS AND MAD MARKETS

·

Ice-Nine and the Non-Darwinian Evolution of Man and Disease

KURT VONNEGUT HAS always been a problem. Is he mere-ly a science fiction writer who happens to write far bet-ter than most of his colleagues? One who also adds just the right level of philosophical trappings to attract a reading audience that normally avoids sci-fi as if it were a virus causing overt idiocy? Or is he a great, thoughtful novelist who uses science fiction as background much as Joseph Conrad used the South Seas or Melville a whaling voyage? And if he is merely a sci-fi *reader,* then why do only serious readers bother to pay any attention to him? Because, ever since Jules Verne, sci-fi writers have an uncanny ability to predict the future. Specifically, I believe Vonnegut may have foreshadowed a disease mechanism that has developed outside of the controls of Darwinian evolu-tion, and as such, has become a threat to man. Is Vonnegut, then, the "author" of non-Darwinian human evolution?

It all begins with *Cat's Cradle,* a novel by Vonnegut that was first published in 1963 and has remained in print ever since.

The story opens with some tales of Dr. Felix Hoenikker, one of the fathers of the atomic bomb. In addition to the atom bomb, Dr. Hoenikker had been given another problem to solve: how to get rid of the mud that clogs up the trucks, landing vehicles, and tanks that the U.S. Marines needed to keep the work safe for democracy. Dr. Hoenikker's solution was a new form of ice, more precisely, a new crystalline structure, one that does not exist in nature. This crystal was a solid at temperatures far above zero and served as a template for all other water molecules that come into contact with it, resulting in a nonnuclear chain reaction. All a Marine would have to do would be to put one molecule into the mud and, presto—no more mud. The problem of course is that all of the oceans and rivers in the world are in touch with each other, so that one molecule can freeze up the entire world. Dr. Hoenikker called this crystalline arrangement ice-nine.

Ice-nine may be the best model we have for describing a set of human and animal diseases that have only been recognized in the last half century, the ones you hear about on the news that seem like something out of the Bible's ten plagues.

And though no one ever thought that it was ice-nine that was destroying the brain of Joseph Ward, I stumbled into a particular theory about this once healthy man. Mr. Ward came to see me, or more correctly, his wife brought him in to see me, after he had been examined extensively back home in Indiana. None of the physicians there had any ideas as to why his brain was seemingly being destroyed at an alarming rate. Joseph was a farmer who raised mostly corn. Until just before Christmas, he'd always been in good health. Then his wife noticed that he seemed to be stumbling, both over his words and his feet.

Joseph said nothing to me, not even hello. His wife did all the talking while he sat in his chair in my examining room, motionless, his face devoid of expression frozen in place. He stared straight ahead, and when I called his name, he didn't respond. I reached out and slowly lifted his right arm. It was slightly rigid, but otherwise resisted being moved in a normal way. I gently put it back in place. As I turned back toward his wife, I dropped a copy of the *Physician's Desk Reference* at his feet, the heaviest book in my examining room. A sudden massive jerk shook his entire body. It lasted for less than a second. His wife nodded to me. She had obviously seen such jerks before. They no longer surprised or frightened her. And it had no effect at all on Joseph Ward. As soon as the jerk passed, he sat there mute and motionless with the same flat expression frozen on his face.

His brain seemed to be dying. But I needed more details.

"How was he, say, around Thanksgiving?" I asked.

"He was himself. We had the whole family over. I always do," she said rather proudly.

"Are you certain?"

"What do you mean?"

"Did you notice anything?" I asked, trying not to lead her on too much, not wanting to create any bias that might result in a factitious history.

"Like what?" She needed to be led.

"Was his appetite good?"

"Fine."

"His mood?"

"He was fine. He loved seeing all of the grandchildren. Always did. He played with them for hours."

"Didn't that tire him out?" Excessive fatigability is common in many neurological disorders.

"Just the opposite."

I pushed her for details. He seemed to need less and less sleep. At first it was just that he didn't seem tired. Then he could hardly sleep at all. He'd never had insomnia before. And when he did fall asleep, he'd have sudden jerks of his body and kick out.

"Like the jerk we just saw?"

"Pretty much."

I wrote my first note on his chart: "insomnia and myoclonus—both spontaneous during sleep and now startle-induced." "Myoclonus" is a ten-cent word for sudden jerks involving many muscles.

Other than his progressive insomnia and some nocturnal myoclonus, she hadn't seen anything unusual until Christmas. Then she noticed he had started to stumble, almost as if he'd had too much to drink. It was not as if he were so drunk that he'd fall down, but he became wobbly.

I thought of the Hermit of Thief River Falls. "Ataxia," I wrote on his chart.

Mr. Ward also started to walk more slowly and shuffle his feet. Mrs. Ward thought that was just to keep himself from falling. I didn't. Someone might intentionally walk more slowly to combat ataxia, but no one ever shuffles his feet to do that. And besides, he had a masklike, frozen expression on his face, and rigidity when I tried to move his arm. All of these—slowness, shuffling, loss of facial movements, rigidity—are part of a Parkinson's-like state. I wrote down "parkinsonism," knowing that Mr. Ward did not have true Parkinson's disease, but merely some of the same manifestations. The course of his disease was not like that of most neurological diseases, in which progression consists of increasing problems in a single system. PD in particular exhibits increasing motor disabilities due to increasing loss of function in a single system. For example, slowness in getting out of a chair progresses to an inability to get out of the chair at all. The changes in Alzheimer's disease from mild memory loss to a state of total loss of cognitive function demonstrate a similar progressive loss of function in one realm only. Even Huntington's disease, with its procession of movement disabilities and dementia, is still restricted to just two systems.

Mr. Ward's history had already provided evidence of the involvement of many systems. The insomnia pointed to alterations in the activating system of the brain stem, the jolts provided involvement of other brain stem structures, the ataxia to the cerebellum and the parkinsonism to the motor systems deep within each cerebral hemisphere. There wasn't much of his brain that wasn't involved. The only major anatomical area that had been spared in the first month was the cerebral cortex itself. That was significant because there is a class of diseases that spare the cerebral cortex and its functions such as speech, and involve only the structures below the cortex itself.

However, by New Year's Day that too had changed. Mr. Ward had begun to stumble over his words, as if he was not able to find the right word, and he often used the wrong word. Together, these two problems—word-finding difficulty and word substitution (or paraphasia)—are part of a progressive aphasia and point to disease of the cerebral cortex itself. So much for sparing the cerebral cortex.

And then suddenly, I thought I knew what Mr. Ward was suffering from: Creutzfeldt-Jakob Disease, or CJD.

As I examined him, my mind wandered more than it should have. CJD is part of a neurological detective story that has intrigued me since I was a resident thirty years earlier. The disease has moved from the verdant pastures of Iceland to the world's last Stone Age population in New Guinea. It has resulted in at least one Nobel Prize in medicine, screaming headlines on the front pages of every newspaper in the world, and is becoming the latest candidate for the plague to end all plagues, the true apocalypse of the third millennium. Is it an example of ice-nine in the brains of humans?

It all started with an Icelandic veterinarian named Björn Sigurdsson, who became interested in two diseases of Icelandic sheep, which were called by their traditional Icelandic names, visna and scrapie. Visna had first been observed in Iceland in the late 1930s and early 1940s and was thought to be related to the arrival of a flock of sheep from Germany in 1933, an interesting year to leave Germany.

The signs of visna usually appeared in fairly young sheep, animals under two years. These early clinical signs included stumbling and lagging behind the rest of the flock. In the natural world of struggle for survival such sheep would have been the first to be eaten, but sheep owners make certain that their livestock do not have to struggle for survival. That is why there are sheepdogs. For sheep stricken with visna, the early slowness and stumbling were followed by paralysis of the hind legs, trembling of the lips, tilting of the head, and finally, over a period of months, total paralysis. The disease was inexorable in its progression: spontaneous remissions or survival into old age were never seen. Pathological studies performed on brains after death showed that the brain cells had all but been wiped out, and there was some evidence of inflammation.

Had the German sheep introduced new genes into the Icelandic sheep stock? The presence of inflammatory changes in the brain suggested otherwise. Such changes are the hallmark of infections, not hereditary "degenerative diseases," which usually lead to deterioration of the brain without any evidence of inflammation.

A second sign of infections is that they are communicable. The chance of spread may be high, as in influenza, or low, as in leprosy (excuse me, Hansen's disease). Visna was slow, progressive, and chronic. Could it be an infectious disease? Attempts to isolate the infectious agent began, but the results were disappointing. The research scientists found no bacteria, no fungi, no parasites. That left only viruses as possible agents, but these were known only to cause acute diseases, and visna was anything but acute.

The question was answered in the early 1950s. Parts of a brain of a single sheep that had visna were inoculated into the brains of healthy sheep. At first, nothing happened. The inoculated sheep remained healthy. One, two, six months—no infection ever studied had taken this long to develop. But the investigators refused to call it quits, and after a year, the once healthy sheep started stumbling and lagging behind the others. Their illness progressed to trembling, paralysis, and eventually death.

Study of their brains confirmed that they had visna, and so it had been proven that visna was infectious and could be transmitted from animal to animal, but only in sheep. All attempts to transmit visna to other animals were unsuccessful. Further studies of the experimentally transmitted disease showed that the sheep can appear to be normal for a far longer incubation period: up to four years can elapse before the signs of paralytic disease develop.

What caused this disease? None of these studies could answer this key question, except to conclude, at least, that the German flock had introduced not a bad gene but a bad virus.

And what about scrapie? The name "scrapie" refers to a persistent tendency of the sheep to scratch or to rub their bodies against trees or fences, which leads to a characteristic patchiness of the wool in the affected sheep. Scrapie was not a new disease in the Icelandic flocks, nor was this disease found only in Iceland: it had been recognized by sheep farmers and veterinarians in many parts of Europe for over two hundred years. Like visna, scrapie is a condition of adult sheep that is invariably fatal and that always follows a chronic course lasting months or even years: progressive ataxia, tremors, and increased excitability. In the end, the sheep become weak and wasted; excessive thirst and blindness are also common.

The early epidemiological studies suggested that scrapie was hereditary. Curiously enough, those sheep who later developed the disease were often unusually well developed as young adults and tended to be used as breeding stock, thereby increasing the chance for transmitting a possible genetic trait. This would, of course, be due to a process of selection not subject to the usual Darwinian constraints. Despite the apparent genetic origin of the disease, veterinary scientists had been able to transmit scrapie to other sheep as early as 1936 by injecting extracts of the brain and spinal cord of affected sheep into healthy ones.

The incubation period of scrapie is quite long, anywhere from nine months to four years. Shades of visna. And, as in visna, the only consistent pathological changes are in the brain and spinal cord. But there is one key difference between the two diseases. The pathologists who studied the brains of sheep with scrapie were never able to find any evidence of inflammation at all. There were no pockets of white blood cells. No traces of any infection. All they found were dead neurons filled with tiny vacuoles, a condition they labeled "spongiform encephalopathy," which simply means spongelike brain degeneration. Still, even without the hallmark inflammatory changes to the brain, scrapie, like visna, was transmissible, had a long incubation period, and was assumed to be caused by a virus.

In 1954, Björn Sigurdsson coined the term "slow viruses" to describe the types of organisms that caused these illnesses in sheep. In so doing, he was clearly differentiating between "chronic" and "slow," two terms medicine has tended to commingle. "Chronic," he felt, had the connotation of a clinical course for the disease that not only was protracted or slow but also tended to be irregular and unpredictable, as is seen in tuberculosis, syphilis, and malaria. In contrast, the sheep diseases investigated by Sigurdsson had a much more predictable course, which he likened to a "slow-motion picture of the chain of events occurring in the acute infection." He therefore proposed the following criteria for slow infections:

1. They have a long incubation period.
2. After clinical signs appear, there is a predictable protracted course, usually ending in death.
3. The infection is limited to a single species; that is, there is only one natural host.
4. The anatomical lesions are found in only a single organ or tissue system within that species.

Sigurdsson did add that "these last statements may have to be modified as knowledge increases." He believed these infections were all caused by viruses. We now know that he was wrong in that assertion and that many of them are caused by peculiar forms of proteins known as prions. But that is getting ahead of the story.

Although the concept of a "slow" infectious agent leading to death from a "degenerative" brain disease was revolutionary, its relationship, if any, to human disease was unclear. Slow viruses were a problem for sheep and veterinarians, not for people and their physicians. That human part of the story began at the opposite end of the world, in what is now Papua New Guinea. It was here that the first proven "slow" infection of the human brain was revealed.

Kuru was first described in 1957. After hearing reports of

this strange, highly localized disorder, scientists D. Carlton
Gajdusek and Vincent Zigas traveled into the Fore tribal areas
of Papua New Guinea to begin an intensive study of it. They, of
course, did not discover the disease, anymore than Columbus
discovered the "New World." Like Columbus they introduced
it to the outside world, in this case to the world of twentieth-
century medicine and science. Kuru existed in a very restricted
mountainous area of Papua New Guinea encompassing about a
thousand square miles and inhabited by approximately 35,000
Melanesian people who were still living in the Old Stone Age,
the prototype of what we once called "primitive." The Fore-
speaking natives, who numbered about 11,000 in all, had the
highest incidence of disease; eight neighboring linguistic groups
were less frequently affected. At that time, the natives of these
areas still lived in small barricaded villages and engaged in tribal warfare, cannibalism, and sorcery. Kuru was well recognized
by the "primitive natives" as a specific disease state. They knew
it was invariably progressive, always led to death, and was the
most common cause of death among the Fore, being primarily
a disease of women and of boys and girls over age five. Only
rarely were adult men affected. More than 1,400 deaths from
kuru were recorded during the first seven years of the study.
This represented over 12 percent of the entire population. Since
kuru was thought by the natives to result from sorcery, ritual
killing of adult male sorcerers added a secondary form of kuru-
related mortality that partially compensated for the severe over-
abundance of men in a population in which kuru had made
women scarce.

The disease started insidiously with tremulousness of the
head and mild difficulty with balance. The Fore people were
clearly far more astute at making an early diagnosis than any of
the Western physicians were. A woman in late pregnancy who
was unable to walk easily across a narrow tree trunk bridging
a gorge knew from that change in her balance that she had kuru
and that she would die of it. The physicians examined her and
thought she was normal, but in less than one year, she was

dead. Her mild imbalance progressed into a frank ataxia (severe imbalance), which followed a relentless course, until she was unable to walk at all; finally she was unable to make the slightest movement without wild tremors. No movement was spared. Slurred speech developed, leading finally to a total inability to speak. What was happening to her, astounding as it was to the trained physicians, was nothing new to the tribe. Shades of Sigurdsson, scrapie and visna. Shades of sorcery. Had she rejected the advances of some sorcerer? Or was she carrying his unwanted child? And, like all other victims of kuru, she starved to death, for ultimately the patient can't swallow.

The pathological abnormalities in kuru were limited to the central nervous system—another of the Sigurdsson criteria fulfilled. On visual inspection, the brains of kuru victims all appeared to be normal. But microscopic examination revealed a severe loss of neurons associated with tiny holes or cavities within the cells, a state called vacuolization. Again, the results suggested a spongiform degeneration, and there was no evidence of infection.

About two years later, Dr. William Hadlow, a veterinarian working with scrapie, pointed out some remarkable similarities between scrapie and kuru in a letter to the editor of *Lancet*. He noted that the epidemiology of the two disorders was similar; each disease was rampant within a confined population; individuals from the isolated populations would first become sick months or even years after leaving the flock or tribe; the disease was introduced into neighboring population groups—by intermarriage, in the case of kuru, or by importation of a ewe or ram, in the case of scrapie.

The clinical signs were also similar. Both kuru and scrapie had insidious courses, manifested primarily by staggering, and both led to death in three to six months. Normal spinal fluid was found in both disorders. And most important, the pathological changes were virtually indistinguishable: neuronal degeneration with vacuolization. Since scrapie had been transmitted by inoculation and was therefore thought to be infectious, Hadlow suggested that kuru might also be a transmissible disorder and that studies in primates should be undertaken to test this hypothesis.

As a direct result of this suggestion, Gajdusek and Clarence Gibbs inoculated chimpanzees with material derived from the brains of Fore natives who had died of kuru. They injected and then they waited: weeks, then months, then years. Eventually, eight chimpanzees that were originally inoculated with kuru material developed ataxia following an incubation period of eighteen months to four years. The imbalance and movement difficulties in the chimps worsened steadily over a four- to six-month period, leading to a moribund state before the animals were finally euthanized. Gajdusek and his associates also reproduced the disease in a second "generation" of chimpanzees by inoculating the brain suspensions of affected chimpanzees.

The gap had been bridged. A degenerative disease in humans had been proved also to be infectious. This was a revolutionary concept. Degenerative diseases had always been regarded as chronic illnesses of deterioration and were therefore neither treatable nor preventable. Nor could they be transmitted. Transmission is a two-way street. Degenerative disease, it was believed, could neither be spread nor caught.

So how did members of the Fore tribe become infected? The mechanism was first suggested by Dr. Robert Glasse, a social anthropologist who lived among the Fore people. Glasse collected evidence that ritual cannibalism had been taken up by the Fore tribe about sixty years earlier, with kuru first appearing about a decade after that. Furthermore, cannibalism among the Fore was practiced primarily by the women, and, in most cases, only the women ate brain tissue of their dead relatives or friends. Women, however, shared these treats with their children, implicating a relationship between the incidence of kuru and the age and gender of Fore members who cannibalized kuru victims. Since the body was consumed with minimal cooking, transmission of an infectious agent would be possible. This mode of transmission could also explain the presence but lower incidence of kuru in surrounding tribes, since the neighboring peoples, though they were also cannibals, refrained from eating known kuru victims.

Cannibalism has been actively suppressed in the Fore area since it came under Western control. As a result there has been a striking decline and finally a total disappearance of the disease. Civilization does have some advantages. Thus a degenerative disease was found to be viral and was eradicated by effective disease control.

What did any of this have to do with my patient, Joseph Ward?

In some ways, Creutzfeldt-Jakob disease is the opposite of kuru. Rather than occurring commonly in a peculiar, isolated population, it is a rare illness that strikes every once in a while throughout the "civilized" world. It occurs in older adults as a progressive dementia reminiscent of Alzheimer's disease, but it is more rapidly progressive and is usually associated with lightning-like muscle jerks (myoclonus). Occasionally it occurs as progressive imbalance. The greatest choreographer of the twentieth century, George Balanchine, died of CJD, which in him started as progressive imbalance.

There is one similarity between CJD and kuru—their pathology. Both are characterized by spongiform changes: the brains of patients with CJD look like those of patients with kuru, and like the brains of sheep with scrapie and visna. Immediately after his success with kuru, Gajdusek and his coworkers carried out the same research protocol on CJD—with the same results. Creutzfeldt-Jakob disease was transmitted to the chimpanzees. At autopsy, spongiform encephalopathy was found. Further studies showed that the disease could be transmitted from chimpanzee to chimpanzee.

This meant that Creutzfeldt-Jakob disease, a degenerative disease, isn't so degenerative after all; instead it is transmissible and infectious. It can be spread to chimpanzees and therefore perhaps to humans. One patient who received a cornea from a patient with CJD developed CJD one year later. A neurosur-

geon who inadvertently cut his hand while doing a brain biopsy on a patient with the disease developed dementia and died in
fewer than two years.

The most alarming part of this story, and the one that had the
greatest implication for Mr. Ward, involved another part of the
body—the pituitary gland. This master gland of the endocrine
system is housed at the base of the skull just below the brain,
and serves as the interface between the brain and the endocrine
system. The pituitary gland develops from the tissue of the
brain. So what? No one eats pituitary glands. That is true. But
for a period of almost ten years we gave children with growth
disorders ("pituitary dwarfs") sufficient amounts of pituitary
hormones obtained from human pituitary glands to help them
grow. Some 10,000 human pituitary glands were ground up to
make each batch of human growth hormone. And some of
those human pituitaries came from patients with CJD or from
patients who had the "virus" for CJD and were still incubating
the disease. How many? Enough that over a dozen children and
adolescents receiving injections of human growth hormone to
overcome pituitary dwarfism died of CJD. And enough was far
too many.

"YOUR HUSBAND HAS what we call Creutzfeldt-Jakob disease," I told Mrs. Ward.

She looked at me as if I were speaking some language she had
never heard. And I was.

"It's caused by a prion."

"Prion," she repeated without any sign of recognition.

What could I tell Mrs. Ward about CJD so that she would
not waste time and money seeking help when there was none?
Treatment is not the only thing we owe patients and their families. Nor even just treatment and compassion. Truth is also a
part of the deal.

I explained that he had a rapidly progressive disease of the
brain for which there is no treatment. She had trouble understanding, or maybe it was denial. This was not the time or place
for long stories about Iceland and New Guinea. She needed
something far less exotic.

Then I made a mistake. I used the only analogy I had left. I told Mrs. Ward about mad cow disease.

She and her husband had been farmers for many years, so I assumed mad cow disease was something they knew about. The technical name for it is bovine spongiform encephalopathy, an infectious degeneration in the brains of cows. It is caused by a peculiar form of infectious or transmittable particle that was given the name "prion" by neurologist and virologist Stanley Prusiner, who thus became the world's first prionologist. In discussing prions and their role in causing diseases, Prusiner described an entirely new agent in disease causation: a protein. ("Prion" is a condensation of "*pr*otein" and "*i*nfectious" plus-*on,* meaning entity.) To date, prions have been implicated in the causation of over half a dozen diseases in animals and/or humans. These include our old friends scrapie, kuru, and CJD, as well as mad cow disease.

All of these infections fit Sigurdsson's original criteria for a slow infection. So of course they are all transmissible. But they can also be inherited. There is a gene for the prion protein, and that gene is found in humans, rodents, and a variety of other species.

What does all this mean? A gene for an abnormal protein which is also infectious? Aren't those two concepts of epidemiology—hereditary versus infectious—all but mutually exclusive?

Most cases of CJD are sporadically distributed. It never comes in epidemics. There are no plagues of CJD. Just a case here and a case there, about one per million per year in the world. And yet 15 percent of cases travel through the generations of a family. It seems likely that prions are ubiquitous in our food chain and yet—and this should prevent any panic over our food supply—the likelihood of prions causing infection is quite low. Clinical manifestations of CJD are far more likely to show up in people who have the prion gene already than in those getting the infection from food chain transmission. Because of these interlocking factors, CJD can be degenerative, transmissible, and familial. A triple-header. And despite the fact that there is a specific prion gene, it has yet to be established the

CJD can arise purely endogenously, that is, on its own from within the body without exposure.

The best example of all this is those few instances of CJD caused by injections of human growth hormone. *All* of the patients receiving the injections had prion proteins in their brains—the prions had been transmitted to all—but only a small number of those who were injected developed CJD. Others receiving injections from the same batches did not develop the disease. So mere exposure to the prion is usually not enough.

But every chimp that had been injected with CJD developed CJD. Did that mean they all had the defective prion gene? No, because the means of transmission is non-Darwinian. And like injection, ritual cannibalization is not a natural factor in the survival of the fittest. In a natural setting, prior to the Stone Age, a man or woman with imbalance would undoubtedly be eaten—but by carnivores, not by humans. So what happened in Papua New Guinea? Did a single case of CJD crop up in a population that practiced cannibalism on themselves? A single case that was then transmitted? And do these small tribal populations provide an especially permissive genetic background?

Back to basic genetics. There is a gene corresponding to Huntington's disease. That gene locus exists in everyone and produces a product. In patients with Huntington's disease, both the gene and the protein product it codes for are abnormal. Although we think in absolutes by considering both the gene and the product to be qualitatively abnormal—and in the clinical sense they are—in reality it is more a *quantitative* difference. That locus is always there. So in exposures like direct brain inoculation there is some sort of focus for the prion to meet up with. In cannibalism, too, with its massive exposure to fresh prions, there is always some focus. But some focuses are more equal than others.

How do these proteins reproduce themselves? They have to do so in order to be transmitted in the laboratory from one animal to another, from generation to generation. If they did not reproduce, the original concentrations of prions would become so diluted as to leave most innoculant solutions free of prion. That was one of the main issues in showing transmission to second and third generations.

There are at least two possibilities for prion replication, one of which uses the cell's DNA in a manner similar to viruses, the other of which doesn't. For the first, remember the usual chain of command: DNA is encoded to make RNA, which in turn produces protein. Not the other way around. However, prions might follow the model of HIV, the AIDS virus. HIV is an RNA virus that, with the aid of two enzymes (together called reverse transcriptase), synthesizes the formation of DNA. Thus HIV can make DNA from RNA—a reversal of the natural order. RNA to DNA to production of infective viral RNA. So the analogous pathway in kuru might be: prion to DNA (that is, to the locus on the prion gene) to RNA to the production of prion molecules. And this gene is everywhere: in humans, mice, sheep (no surprise there), goats, nematodes, fruit flies, even yeast—all sources of prions for all of these related diseases.

And the other scenario? Carleton Gajdusek wrote a paper entitled "Fantasy of a 'Virus' from the Inorganic World." In it he hypothesized an infectious particle acting more like a crystal than an organic species. Shades of Vonnegut and ice-nine: physical-chemical reproduction, neither sexual nor asexual. Merely recruitment of the right amino acids in the right order. A protein acting as its own template and "reproducing" through polymerization—by adding more and more of its own molecules onto itself.

But why does this ice-nine-like activity cause only neurological disease? Why not liver disease or kidney failure?

It turns out that neurons within the brain are virtually the only cells in the human or other mammalian species that synthesize prions. It's as if Vonnegut was wrong: ice-nine works for only certain waters, in mud but not in the oceans.

There are any number of possible explanations for this. The brain is immunologically isolated. Perhaps other areas of the body have defenses that prevent prion replication. Or maybe it's the unique nature of neurons that's important. Unlike other kinds of cells, neurons have lost the ability to reproduce, though they've maintained the ability to survive. We die with the same nerve cells we grew up with. No other organ works like that. How can a neuron live for a hundred years? By continually repairing itself, replacing its parts: membranes, enzymes, its very structure. A neuron is always making structural proteins and replacing old proteins with new ones. So when a prion enters the cell, ready to be reproduced—in a machine geared to reproduce proteins, with a rich pool of amino acids—it's like a kid walking into a candy store.

So new prion proteins are made. And they polymerize. And then they congregate. And they form deposits, which, like ice-nine, gum up the works, eventually injuring the membranes that are due to be replaced. And in so doing the prions kill the nerve and are released into the brain. Just where they want to be: to be taken up into another nerve cell so the process can start all over again.

This was what was going on inside Joseph Ward's brain, or to be more specific, inside the individual neurons within his brain. But he did not have bovine spongiform encephalopathy.

MAD COW DISEASE first appeared in the United Kingdom in 1986. It was a new disease for British cattle. The cattle would begin to stagger and behave strangely, and then within months weaken and die. When studied the disease looked a lot like scrapie. Back to Sigurdsson. Did it fill all of his criteria? It certainly did. And the infected brains contained prions.

Where did those prions come from? Had the scrapie prion somehow learned to cross the species barrier after all these centuries? Or was there some other source? And what source could there be in ruminants that were strict vegetarians? Or were they pure vegetarians? The cattle all thought that they were. That's the niche in the environment they evolved to fill. Cows are strict herbivores. Their teeth are only fit for grinding plants, and for rechewing their regurgitated cud. Their long gastrointestinal tracts must work for the prolonged time it takes to break down cellulose and derive the needed nutrition from inside the plant cells.

But man, in his infinite wisdom, has changed all of that. Grass is not a very good source of calories or minerals. And corn is so expensive. But rendered meat is cheap. It is full of protein, has all those amino acids that cows need to build up more cow. And there is lots of calcium too. Rendered meat is just what it is advertised to be: a protein- and calcium-rich slaughterhouse by-product that is produced by grinding together whatever cannot otherwise be used from any animal that goes through the slaughterhouse. This includes all animals brought to market to be slaughtered and thereby enter the human food chain (a process that all but assures survival of our species). But rendering also includes animals not intended for the human food chain. Animals that have died for whatever reason, including sick cows, and—you've guessed it—sick sheep from herds with scrapie. That is, sheep from herds known to have scrapie, sheep who may well have had the prion in their bodies but still seemed quite normal. The infectious prions are not in the ground-up muscles of the dead animals, the main source of proteins and amino acids, nor in the bones, which are used as a source of much-needed calcium. The prions are where they always are—in the nervous system, in the brains and the spinal cords. The latter is important. It is easy to discard the brain during processing, but when vertebrae are made into bonemeal, the spinal cord is left in because it's too hard to take out. The British cows of 1986 had, in a sense, become omnivorous. They'd even become cannibals.

It's a scenario that Darwin never envisaged, one for which the cows had no defense, anymore than those chimps in Gajdusek's lab did. But this time there was no Nobel Prize waiting in the wings. By changing the pattern of bovine food intake, we created a new disease. The defenseless herds were quickly devastated. It doesn't take much feed to kill a cow; one gram of prion-infested material can do it. Entire herds were infected. Within a few short years, thousands upon thousands of cattle died, and tens of thousands more were killed in an attempt to control the disease.

For a year or two after mad cow disease made its appearance, a new form of CJD appeared in the United Kingdom, a type that was clearly different from plain, old-fashioned CJD. All of the patients were young, mostly adolescents. They did not exhibit the more typical staggering related to CJD. Instead these young people began showing signs of psychiatric problems and changes in the behavior. First they became anxious, then depressed and withdrawn. Next they became even more obviously disturbed, exhibiting schizophrenic-like behaviors. Finally came progressive neurological deterioration and death. Neurologically the patients developed progressive motor problems (often a kuru-like ataxia) a CJD-like dementia, and myoclonus. And all eleven of those patients who died had spongiform encephalopathy.

Did this new condition indicate that mad cow disease was spreading to susceptible humans?

Did Joe have mad cow disease? Mrs. Ward wanted to know.

No, not really. He had Creutzfeldt-Jakob disease, a first cousin. God forbid the authorities should ever admit that. To them it was just two new diseases, both linked to prions and therefore to protein sources in the diet, both just happening to occur at the same time in the same relatively small geographic area and nowhere else in the world either simultaneously or ever before. It was, in their best judgment, merely a coincidence. Nature abhors coincidences.

Was I absolutely certain?

Not absolutely. That would require a brain biopsy.

A what?

A study of his brain.

She didn't want Joe to go through that unless it would help him.

It would not help him. It would merely confirm what we already knew.

"A brain biopsy," she mumbled unenthusiastically.

Or an autopsy, I thought to myself. "A brain-wave test might help," I added. It wouldn't help me. But having something more than just my clinical opinion would help Mrs. Ward. Almost 90 percent of patients with CJD and myoclonus have characteristic EEG discharges specific to the disease.

Mrs. Ward and I watched as the recording pens scratched out a record of Joseph's brain activity. Each time he had one of his jerks, a burst of abnormal activity exploded across the paper followed by nothing, no electrical activity at all. The entire brain ceased to create brain waves. These explosive paroxysms of suppressive activity are called suppression bursts. They are characteristic of old-fashioned CJD. They had never been seen in the new variant of CJD occurring in the U.K. That is, none of the patients suffering from prion disease related to mad cow disease also had suppression bursts. Not even those who developed myoclonus. I explained all of this to her.

Now at least she had an answer. She had seen the abnormal bursts that were destroying her husband's brain. Well, the prions were. And the brain was responding to the destruction by creating suppression bursts.

"And there is nothing you can do for him?"

"There is nothing anyone can do."

"No cure?"

"No cure."

"No treatment?"

"No treatment."

"No nothing?"

"No nothing."

That inarticulate response said all that could be said. Its effect on her was worse than any suppression burst I had ever seen. And the silence that followed was deafening.

She took Joe back home to Indiana, to the farm. He died four weeks later. I heard about it from the local medical examiner, who called me. He was going to do an autopsy, and since Joseph Ward had mad cow disease, he wanted to know what precautions to take.

But Joseph Ward didn't have mad cow disease, I argued.

Mrs. Ward had said that was what the doctors told her. The doctors at the medical school in Chicago.

But I was "those doctors," I protested in vain.

Mrs. Ward had also told the medical examiner exactly how

Joseph had gotten mad cow disease: from the bonemeal he'd used on his beloved rose bushes.

I took a deep breath and started in again.

Joe Ward had died of CJD, I said.

Was that my opinion?

It was my diagnosis. I was the one who said he had prion disease. Without my intervention no one would have thought he had either CJD or mad cow disease. He would have just died of some obscure and undiagnosed disease.

Did I have any tests?

I did. The encephalogram.

And what had that shown?

"Suppression bursts."

That meant nothing to him until I explained. Then he was convinced.

Yet the very next day, a local paper ran the story. Joseph Ward, a farmer, had died of mad cow disease, the first proven case of mad cow disease in the United States. Perhaps it was the beginning of a plague. Or at least an epidemic. The apocalypse, like the millennium, might be just around the corner. It sold newspapers.

The Associated Press picked it up that evening, and the next day the bottom fell out of cattle, corn, and soybean futures. Both corn and soybeans are used to feed cattle. Cattle futures tumbled the daily allowed limit within minutes.

And all of this occurred because I had tried to explain to Mrs. Ward what was happening to her husband. The Indiana Department of Public Health, the Department of Agriculture, and the Centers for Disease Control and Prevention in Atlanta scrambled to set the record straight. Reporters even bothered to interview the physician who had made the diagnosis of mad cow disease in the first place, or at least that was how most of the reporters seemed to identify me.

In a couple of days the markets recovered. I am not certain all of the individual investors ever did. Many of my friends wanted me to let them know the next time I made a diagnosis of mad cow disease so they could go short on cattle futures. I protested I had never made the diagnosis in the first place. I had never seen a patient with mad cow disease. I probably never would.

Next time, tell us first, they repeated.

There are many lessons to be learned here. One is to be careful when using analogies.

More important, though, this story in essence is all about evolution. Evolution is made up of a continuing series of small, not-quite-random changes in chromosomal DNA which produce equally small and equally not-quite-random alterations in the structure or function of those members of the succeeding generation that inherit those minor alterations. Such alterations in turn influence the ability of the not quite randomly altered offspring to compete for reproductive survival. It was—and still is when not tampered with—a slow, arduous process without any built-in sense of direction. Man has changed all of that.

For better and for worse, we created mad cow disease. Oh, we didn't create the prions. They have probably been around forever. So has the gene for the prion protein, which must have been around since the age of the dinosaurs, if not before. (It turns out not to be in all species. Pigs do not have the gene, for example. "That other white meat" may be safer than even the advertisers claim.) Until the creation of inbred, genetically isolated herds of sheep, in which reproductive fitness was determined by humans, not the environment, the viral visna and (prionic) scrapie couldn't be passed along.

Until an isolated tribe took up cannibalism on a relative, who had lived long enough to make enough prions to have what I would call "sporadic CJD," kuru didn't exist. Here again, man had intervened.

Cows weren't turning the people who ate their meat mad until we turned the cows into omnivores and fed them their dead relatives, so to speak. Shades of Sigurdsson and Vonnegut.

But these are not the only lessons. All recent human "evolution" and development now occurs without the participation of DNA. In fact, almost all of it takes place outside the body. This is because we can alter our environment in ways no other species ever could and our self-shaped environment plays a crucial role in the postnatal development of the human brain. Our genetic makeup today is not significantly different from that of the caveman fifty thousand years ago, yet nothing else about our lives is similar. That sort of change is unprecedented, could not have been predicted by Darwinian theory, and is unrelated to any of those not-so-random genetic variations in the DNA of our chromosomes. Unfortunately this form of evolution is carried on without the mechanism of survival of the fittest, a form of feedback we have yet to replace in any meaningful way.

We have survived in ways and numbers that could not have resulted merely from the slow evolution of DNA. We do not run very well. We are not really very strong. Our sense of smell is abysmal, our hearing is limited, and our vision is third-rate. Yet we threaten the very survival of every species that is faster, stronger, and has better-developed special senses. This is because of what our brains are able to do in between sensory input and motor output. What is perplexing is how this all came about and why the brain has evolved into a mechanism for learning and adaptation that far outstrips whatever simple tasks it was biologically evolved to perform. A vocabulary of a few hundred words would have been sufficient to produce the biological advantage needed to assure survival of the species. Yet the average educated American can read over one hundred thousand words, uses over ten thousand words in his spoken language, and learns hundreds of new words each year—from "gigabyte" to "Prozac" to "slam dunk"—without even thinking about the process. And each human infant learns to generate sentences and ideas using these words long before he or she learns to control bladder and bowel function.

Back to New Guinea. Back to the Fore and the study of kuru.

How long had kuru been plaguing these people? Was it a new disease, a plague that had recently arrived like the Spanish flu and would inevitably run its course, or had it always been there? The tribal elders were asked. Their answer was "many" generations. It was assumed this meant a very long time. That eliminated cannibalism as a means of spreading the disease, since cannibalism had been introduced less than sixty years earlier.

Then a couple of anthropologists (Robert Glasse and Shirley Lindenbaum) came along and analyzed the North Fore language. The counting system of these tribes consisted entirely of one, two, three, and many. Their "many" included every quantity from a few all the way to infinity. Cannibalism was once again considered the means of spread of kuru. With the suppression of cannibalism, the disease has been eliminated.

Understanding the North Fore language not only was a pivotal step in preventing kuru; it also sheds light on the progressive nature of human intellectual development. My granddaughter just passed her second birthday. To me she is, of course, both beautiful and brilliant. That goes without saying. And very verbal. She knows one, two, three, four, and five—not just as words, but as symbolic concepts. She can differentiate "four" from "three," and "five" from "four," and "many" from all of those numbers that her brain has integrated into its circuitry. The Fore can only count to "many." But my granddaughter's mind, given the proper environmental exposure, has in two years outstripped everything that humans struggled 5 million years to attain for the first time. The means for her evolution is not, in this case, the slow arduous changing of DNA, through survival of the the fittest, but rather, her environmental exposure—non-Darwinian evolution.

Our future evolution is no longer tied to DNA. No more than our recent past has been. We, that is, those of us who live in developed countries are flourishing for cultural reasons, and cultural reasons alone. In a few short decades, sewage has been isolated from the water supply, the average life expectancy has doubled, infant mortality has declined, and survival to reproductive fitness has challenged the food supply. There's always an up- and a downside.

One of England's most famous Stone Age hunter-gatherers is known today as Cheddar Man. When he died some nine thousand years ago, he was buried in a cave in what is now Cheddar Gorge in southwest England. All of his fellow hunter-gatherers were recently in Cheddar Gorge, participating in the filming of the local history of the town, one of those interminable BBC documentaries. The producer asked a simple question: "Are there any surviving descendants of Cheddar Man still living in Cheddar?"

What a great notion!

Researchers from Oxford University and the London Museum of Natural History set out to answer that question. They started by extracting DNA from a tooth of Cheddar Man and determining the sequence of base pairs in a section of that DNA. After sequencing Cheddar Man, they then collected samples of present-day citizens of Cheddar Gorge. And they found a match. Cheddar Man is living on in the guise of a direct descendant named Adrian Targett, who teaches modern history at the Kings of Wessex Community School.

So the DNA hasn't changed much in Cheddar.

But nothing else is the same.

Cheddar Man's direct descendant teaches in a school that his forebear could never even have conceptualized. Now that's progress. And what's more, the Fore no longer count "one, two, three, many." They, too, have left the Stone Age behind them. It has not even taken them one generation to move from the Stone Age to the information age, much less than the ninety centuries it took Cheddar Man and his descendants.

1 3

WHATEVER HAPPENED TO
BABY NEANDERTHAL?

·

An Afterthought

THIS AFTERTHOUGHT REVISITS some observations on our traditional model of the ascent of modern man. According to this model, the tree of ascent, Neanderthal man branched off into an isolated dead end half a million years ago or even earlier, and the family of modern man continued to evolve along another pathway, finally dividing into separate "races." This theory is based on the nature of human mitochondrial DNA. The thirty-seven genes in mitochondrial DNA vary only a little from individual to individual, from one "race" to another, from black Africans to Native Americans, to Asiatics, to Pacific Islanders. It's also true of people from Stone Age peoples, such as the Fore population in Papua New Guinea, to the information-age pioneers: very few nucleotides within our mitochondrial DNA are any different.

This has been interpreted to mean that we are all descended from the same mother (or small group of mothers) only 200,000 years ago.

Was there just one Eve? And why only Eve, and not Adam and Eve?

We think of all genetic material and all genetic heritage as residing within the chromosomes and being inherited randomly from both parents. There is one exception to this rule: the mitochondria, the site of oxidative metabolism (or energy production), have their own DNA. It is as if the mitochondria were at one time bacteria-like parasites that wandered into a single-celled organism and have lived there ever since. Eggs and sperm, like all other human cells, have mitochondria. But only the *nucleus* of the sperm, which contains chromosomal DNA but no mitochondria, enters the egg. So each of us inherits 100 percent of our mitochondrial DNA from our mothers. No paternal mitochondrial DNA lives on. And if all maternally derived DNA is so similar, then there really must have been only one mother of us all.

And why only 200,000 years ago?

It is possible that a single mother or a small group of mothers survived some sort of evolutionary bottleneck, a situation that resulted when the struggle to survive became extremely competitive and only a limited number of potential mothers were able to survive. And those who came through the bottleneck with their reproductive capacity intact did so because their mitochondria differed from the mitochondria of others in a manner that gave them a biological advantage.

Such bottlenecks must have taken place, for without them much of paleontology would be hard to fathom. In the conventional picture of evolution, called gradualism, change is slow and steady, with constant minor alterations in each type of organism adding up to their evolution. If that model were true, however, how could the same species of dinosaur stick around for 10 million years or more and then in a short time evolve into a new species? If evolution consisted of a uniform rate of change, it would be almost impossible to separate out different species. Much of evolution must consist of periods of equilibrium in which the environment is fairly stable. Animals and plants selected to live in such an environment would change little over millions of years. Then something happens, and the environment changes. The balance shifts, different variants become advantageous, and new characteristics become favored. Change! Evolution! The beginning of a new species then thrives relatively unchanged in the new environment until the battle-field (the environment) again changes. Niles Eldridge and Stephen Jay Gould have called this model of evolution based on bottlenecks "punctuated equilibrium."

Traditionally, we believed that Neanderthal man was isolat-ed from the evolution of modern humans: two separate trunks on the evolutionary tree which lived in separate places at the same time or the same place at separate times, but never the twain did meet. Neanderthal then died out later, to be replaced by modern *Homo sapiens*. All well and good, but unfortunate-ly not true. Recent discoveries have demonstrated that Neanderthal man and modern man lived in the same regions (southern Europe, Israel) at the same time, for at least ten thou-sand years. They shared the same technologies and even the same caves. That's pretty close for two separate trunks.

While not wanting to cast any aspersions on the mores or behavior habits of Neanderthals, it is inconceivable that mod-ern man *did not* cohabit with the Neanderthals. And the odds are that such cohabitation would have resulted in plain, old-fashioned biblical begetting. So where are the descendants of such begetting? There aren't any. Neanderthals' DNA has dis-appeared without leaving a single trace in the DNA of *Homo sapiens*.

What do we know about Neanderthals? They had a thick, heavy body with enormous upper-body strength. They were obviously much stronger than modern humans. They had a for-ward-thrust face with a receding chin. Their head also project-ed forward, and the hole in the skull through which the spinal cord exited the cranial cavity—the foramen magnum—was angled very differently than ours. But none of that should have mattered in survival. It must have been the brain that mattered. And theirs was bigger than ours, and their speech box was fully capable of supporting speech. Ergo, they must have had language. Neanderthals spoke, not the ejaculations of monkeys, but true symbolic speech.

Yet their civilization failed so completely that even cohabitation failed to leave a single trace of Neanderthal DNA.

So why did we succeed when they failed?

The brain of the Neanderthal was far bigger than ours right from birth. We know this because of their large pelvic area, documented most recently by the discovery of a complete Neanderthal pelvis in Israel. Thus Neanderthal not only died with a bigger brain; they were born with it too. The latter, as counterintuitive as it may seem, was most likely a major biological disadvantage. Such a brain would have been more completely developed at birth. There was far less need for prolonged postnatal development, with all that complicated interaction with the environment.

In short, the Neanderthal had a larger, more mature brain at birth, which therefore evolved through a far shorter and less significant period of juvenilization. Recall the difference between humans and chimpanzees. We probably had a common ancestor some 4 to 5 million years ago. Remember the discussion earlier in this book on juvenilization of the human brain and how that slow development allows for much environmental influence. Most of the development of the human brain takes place outside of the womb, within the real world. And that real world shapes what the brain can do and how well it does it—a distinct biological advantage.

What about baby Neanderthals?

Developmentally they fell somewhere between the chimp and the human: more juvenilization than in chimps but less than in humans, fatally less.

The greater maturation of the Neanderthal brain at birth would have been a biological advantage only so long as survival depended more on a classic survival of the fittest than on the acquisition of knowledge within a human-manipulated environment.

Three factors have been traditionally tied together in our view of the ascent of man: bipedalism (walking upright on two legs), big brains, and toolmaking. It used to be taught that these three traits went hand in hand—no pun intended. Now it is clear that bipedalism came first, by a couple of million years. Why? Because it does more than free the hands to make tools. Kangaroos are bipedal, but they don't make very many tools. That is not the primary advantage of bipedalism. Toolmaking is really the result of a spandrel, one that gave great biological advantage, but still is a spandrel. Bipedalism changed the size and shape of the pelvic cavity. It shrank the dimensions of the outlet in the pelvis. For chimpanzees, childbirth is relatively painless and safe. Not so for bipedal women. Infants develop within the womb who just won't fit through. That's called absolute cephalopelvic disproportion. And if that is still occurring now, some 4 million years after the invention of bipedalism, then it must have been far more common at the beginning. It clearly is an impediment to further reproduction since, before medical technology could manage this problem, both mother and child would always have failed to survive the ordeal.

The biological alternatives were simple. Smaller brains at birth was one option. Smaller brains require smaller heads, and those smaller heads can get by with smaller pelvic outlets. So the brain after birth stays the same size as the brain of a fully developed fetal chimp. A second pathway for evolution is to develop thicker bodies with larger pelvic outlets. That, at least in part, seems to have been the Neanderthal choice. Then there is the third option: having a small chimplike brain at birth but a prolonged period of post-birth dependency with juvenilization of the brain. This was the choice that evolution selected for the direct ancestors of modern man: bipedalism followed by juvenilization of the brain with enhanced brain size and adaptability, which led to toolmaking, language, and all the rest.

How did this sequence of development occur? During a period of punctuated equilibrium, at some bottleneck or other. If all Neanderthal DNA was eliminated from the human gene pool at that time, then there may well have been a relationship between mitochondrial DNA and brain development. And perhaps our maternal mitochondrial DNA played a role.

Diseases of human mitochondrial DNA, such as Leber's hereditary optic atrophy, are always inherited from the mother. Such diseases usually involve two systems of the human body and often cause weakness of the muscles. This is far from surprising since muscles depend so much on oxidative metabolism, which after all is what the mitochondria do best. But these diseases also cause brain dysfunction, which is why they are grouped together under the rubric "mitochondrial encephalopathies"—brain disorders due to the mitochrondria. The existence of these disorders and the way they cause selective brain dysfunction demonstrate the pivotal role mitochrondrial DNA plays in postnatal development of the human brain.

So what?

So everything.

Enter a modern man. He cohabits with a Neanderthal woman. The brains of their offspring could well be half Neanderthal and half human at birth—smaller than a Neanderthal brain, yet unable to develop into human brains. Therefore they are at a major disadvantage, neither fish nor fowl. Moreover, their mitochrondrial DNA is 100 percent Neanderthal. No help there in prolonging juvenilization. The result? Extinction. Without a trace.

Enter a modern woman. She cohabits with a bigger and stronger Neanderthal, whether by choice or not. He contributes no mitochondrial DNA to their offspring. Armed instead with the more advantageous maternal human DNA from their mother, they could experience the process of prolonged juvenilization, but only the "human" mitochondrial DNA would get transmitted to any subsequent generations. But sometimes the brain is too big at birth. The small pelvis that may have made her sexually attractive cannot accommodate that big head, which is halfway in size between the small-birth brain of a chimp or human and the far larger brain of the fully developed fetal Neanderthal. The result is absolute cephalopelvic disproportion and the death of mother and child. This bigger-brain baby does not survive.

But why did Neanderthal leave no trace in the rest of our genetic makeup?

The most important neurological function likely missing from Neanderthal man would have been complex language. While their limited voice box could have been a factor in this, brain function was far more significant. This difference has been overlooked by most paleontologists and even by paleoanthropologists. Christopher Stringer of the London Natural History Museum has recently suggested that Neanderthal man lacked a complex language because his social life and culture did not require one. Such reasoning is teleological, that is, it presumes that the need for language existed before complex symbolic language did, a supposition for which there is no evidence and scant logical basis. It is far more reasonable to conclude that Neanderthal man did not develop a complex language because he couldn't (based on the limitations of his brain) and thereby had a social system that did not demand such a language. Certainly, if language development merely required a need for its development, what greater need could there be than the threat of extinction? If that does not equate to need, I don't know what would. So if the fathers could not contribute maternal mitochondrial DNA, then the offspring couldn't profit from juvenilization. Though stronger and more fierce, the baby Neanderthals were unable to adapt. The evolutionary race goes not to the strong, but to the more adaptable, the more wily, the more juvenilized. In essence, Neanderthals bred themselves out of existence.

Thank God for prolonged adolescence. And Darwin.

Think about that the next time your adolescent demonstrates a total lack of adult judgment. That's how we got here. If it weren't for such juvenile thinking, actions, and behavior so late in life, we might not have survived the Neanderthal! We might still be in those cold caves, cohabiting with Neanderthals instead of enjoying the fruits of our progress from reruns of *Amos and Andy* and remakes of *The Mark of Zorro*. If that isn't progress, what is?

INDEX